55 SECRETS & TRICKS IN MATHEMATICS

© 2014 Davies Guttmann

ISBN 9783735741264

„Herstellung und Verlag: BoD – Books on Demand, Norderstedt"

Bibliografische Information der Deutschen Nationalbibliothek: Die Deutsche Nationalbibliothek verzeichnet diese Publikation in der Deutschen Nationalbibliografie; detaillierte bibliografische Daten sind im Internet über www.dnb.de abrufbar.

Whether we realize it, like it or not mathematics controls our lives.

The time of day, the distances we travel, the money we earn and spend – these are all mathematical constructs that translate into daily living.

But it goes much deeper than that – think of all the machines and equipment we use; the computer you are probably reading this on, the TV set, your car. These inventions all depended on mathematics to make them happen and continue to rely on it for development and improvement.

Similarly think of the frequent transactions you do that are based on mathematics – shopping, filling out your tax return, gambling, working out your mileage expenses – even working out that 10% tip on your restaurant bill. All mathematics related activities.

The truth is that without mathematics the world would not be able to function. You might not like mathematics or not be very good at it – many people are like this – but you cannot escape from it in your lives. And a deficiency in understanding and appreciating it can handicap you in very specific and significant ways.

Overcoming the fear of maths and stripping away some of the mystery and complexity is the best way to start gaining control of how it affects your life. This collection looks at a lot of the tricks and secrets that can make you much more confident and proficient when it comes to numbers – and make sure you are not conned or bamboozled by those who understand the relationships between numbers and real world activities better than you might have previously.

If you dreaded maths at school and have pretty much avoided contact with it ever since this is the book to reintroduce you to the concept in a much more friendly and usable way.

Contents

Shortcuts in Multiplication, Division, Addition
    & Subtraction 1
Mental Math Shortcuts 18
April Fool Math 32
Always 5 34
Five Statistics Problems That Will Change The
    Way You See The World 36
Steps to Doing Well in Math 48
Math Magic/Tricks 54
Is it Magic or Is it Maths? 59
The Calendar Trick 61
The Secret World of Codes and Code Breaking 65
The Secret to Success in Mathematics 72
Maths and magic 78
Pythagorean Mathematics 85
11 Super Badass Math Tricks 103
How to learn math formulas 113
11 Math Tricks That Will Make
    Your Life So Much Easier 119
Techniques for Mental Arithmetic 131
Fast Arithmetic Tips 138
Do Super Quick Maths Calculation Using Vedic Method ... 152
10 Easy Arithmetic Tricks 157
Figures of fun: Amaze your friends with these
    fantastic maths magic tricks 166

Use these calculator tricks to impress and
    astound your friends!... 174
Weird mathematical tricks that shouldn't work - but do 180
Not statistically significant and other statistical tricks. 187
How to Lie and Cheat with Statistics 192
Weird Statistics 202

Shortcuts in Multiplication, Division, Addition & Subtraction

These are some of the practical secrets in mental math calculation speed. Knowing these can help you figure out answers to most of life's daily activities involving numbers. There are many useful Math techniques, tricks, and secrets that can be valuable in your day to day work or study. Knowing these shortcuts are key to computation speed and accuracy in your Mental Mathematical Aptitude. This can be useful in applications such as IQ tests, aptitude tests, job application tests, military tests, college entrance tests and so many more uses in the office, workplace, or in your own home.

SHORTCUTS IN MULTIPLICATION:

Multiplication using multiples
12 x 15
= 12 x 5 x 3
= 60 x 3
= 180

Multiplication by distribution
12 x 17
= (12 x 10) + (12 x 7) ---> 12 is multiplied to both 10 & 7
= 120 + 84
= 204

Multiplication by "giving and taking"
12 x 47
= 12 x (50 - 3)
= (12 x 50) - (12 x 3)
= 600 - 36
= 564

Multiplication by 5 --> take the half(0.5) then multiply by 10
428 x 5
= (428 x 1/2) x 10 = 428 x 0.5 x 10
= 214 x 10
= 2140

Multiplication by 10 ---> just move the decimal point one place to the right
14 x 10
= 140 ---> added one zero

Multiplication by 50 ---> take the half(0.5) then multiply by 100
18 x 50
= (18/2) x 100 = 18 x 0.5 x 100
= 9 x 100
= 900

SECRETS & TRICKS OF MATHEMATICS

Multiplication by 100 ---> move the decimal point two places to the right
42 x 100
= 4200 ---> added two zeroes

Multiplication by 500 ---> take the half(0.5) then multiply by 1000
21 x 500
= 21/2 x 1000
= 10.5 x 1000
= 10500

Multiplication by 25 ---> use the analogy $1 = 4 x 25 cents
25 x 14
= (25 x 10) + (25 x 4) ---> 250 + 100 ---> $2.50 + $1
= 350

Multiplication by 25 ---> divide by 4 then multiply by 100
36 x 25
= (36/4) x 100
= 9 x 100
= 900

Multiplication by 11 if sum of digits is less than 10
72 x 11
= 7_2 ---> the middle term = 7 + 2 = 9
= place the middle term 9 between 7 & 2
= 792

Multiplication by 11 if sum of digits is greater than 10
87 x 11

SECRETS & TRICKS OF MATHEMATICS

= 8_7 ---> the middle term = 8 + 7 = 15
because the middle term is greater than 10, use 5 then add 1 to the first term 8, which leads to the answer of
= 957

Multiplication of 37 by the 3, 6, 9 until 27 series of numbers --> the "triple effect"
solve 37 x 3
multiply 7 by 3 = 21, knowing the last digit (1), just combine two more 1's giving the triple digit answer 111

solve 37 x 9
multiply 7 by 9 = 63, knowing the last digit (3), just combine two more 3's giving the triple digit answer 333
solve 37 x 21
multiply 7 by 21 = 147, knowing the last digit (7), just combine two more 7's giving the triple digit answer 777

Multiplication of the "dozen teens" group of numbers --
(i.e. 12, 13, 14, 15, 16, 17, 18, 19) by ANY of the numbers within the group:
solve 14 x 17
4 x 7 = 28; remember 8, carry 2
14 + 7 = 21
add 21 to whats is carried (2)
giving the result 23
form the answer by combinig 23 to what is remembered (8)
giving the answer 238

Multiplication of numbers ending in 5 with difference of 10
45 x 35

first term = [(4 + 1) x 3] = 15; (4 is the first digit of 45 and 3 is the first digit of 35 --> add 1 to the higher first digit which is 4 in this case, then multiply the result by 3, giving 15)
last term = 75
combining the first term and last term,
= 1575

75 x 85
first term = (8 + 1) x 7 = 63
last term = 75
combining first and last terms,
= 6375

15 x 25
= 375

Multiplication of numbers ending in 5 with the same first terms (square of a number)
25 x 25
first term = (2 + 1) x 2 = 6
last term = 25
answer = 625 ---> square of 25

75 x 75
first term = (7 + 1) x 7 = 56
last term = 25
answer = 5625 ---> 75 squared

SECRETS & TRICKS OF MATHEMATICS

SHORTCUTS IN DIVISION:

Division by parts ---> imagine dividing $874 between two persons
874/2
= 800/2 + 74/2
= 400 + 37
= 437

Division using the factors of the divisor: "double division"
70/14
= (70/7)/2 ---> 7 and 2 are the factors of 14
= 10/2
= 5

Division by using fractions:
132/2
= (100/2 + 32/2) ---> break down into two fractions
= (50 + 16)
= 66

Division by 5 ---> divide by 100 then multiply by 20
1400/5
= (1400/100) x 20
= 14 x 20
= 280

Division by 10 ---> move the decimal point one place to the left
0.5/10
= 0.05 ---> 5% is 50% divided by ten

Division by 50 ---> divide by 100 then multiply by 2
2100/50
= (2100/100) x 2
= 21 x 2
= 42

700/50
= (700/100) x 2
= 7 x 2
= 14

Division by 100 ---> move the decimal point two places to the left
25/100
= 0.25

Division by 500 ---> divide by 100 then multiply by 0.2
17/500
= (17/100) x 0.2
= 0.17 x 0.2
= 0.034

Division by 25 ---> divide by 100 then multiply by 4
500/25
= (500/100) x 4
= 5 x 4
= 20

750/25
= (750/100) x 4
= 7.5 x 2 x 2
= 30

SECRETS & TRICKS OF MATHEMATICS

SHORTCUTS IN ADDITION:

Addition of numbers close to multiples of ten (e.g. 19, 29, 89, 99 etc.)
116 + 39
= 116 + (40 - 1)
= 116 + 40 - 1 ---> add 40 then subtract 1
= 156 - 1
= 155

116 + 97
= 116 + (100 - 3)
= 116 + 100 - 3 ---> add 100 then subtract 3
= 216 - 3
= 213

Addition of decimals
12.5 + 6.25
= (12 + 0.5) + (6 + 0.25)
= 12 + 6 + 0.5 + 0.25 ---> add the integers then the decimals
= 18 + 0.5 + 0.25
= 18.75

SHORTCUTS IN SUBTRACTION:

Subtraction by numbers close to 100, 200, 300, 400, etc.
250 - 96
= 250 - (100 - 4)
= 250 - 100 + 4 ---> subtract 100 then add 4
= 150 + 4
= 154

250 - 196
= 250 - (200 - 4)
= 250 - 200 + 4 ---> subtract 200 then add 4
= 50 + 4
= 54

216 - 61
= 216 - (100 - 39)
= 216 - 100 + 39
= 116 + (40 - 1) ---> now the operation is addition, which is much easier
= 156 - 1
= 155

Subtraction of decimals
47 - 9.9
= 47 - (9 + 0.9) ---> "double subtraction"
= 47 - 9 - 0.9 ---> subtract the integer first then the decimal
= 38 - 0.9
= 37.1

18.3 - 0.8

= 18 + 0.3 - 0.8
= (18 - 0.8) + 0.3 ---> subtract 0.8 from 18 first
= 17.2 + 0.3
= 17.5

WORKING ON PERCENTAGES:

30% of 120
= 10% x 3 x 120 ---> it is much easier working with tens (10%)
= 10% x 120 x 3
= 12 x 3
= 36

five percent of a number: 5%
360 x 5%
= 360 x 10%/2 ---> take the 10% and divide by 2
= 36/2
= 18

360 x 5%
= 360 x 50%/10 ---> take the half(0.5) and divide by 10
= (360/2)/10
= 180/10
= 18

ninety percent of a number: 90%
90% of 700
= (100% - 10%) x 700
= (100% x 700) - (10% x 700) ---> 100% minus 10% of the number
= 700 - 70
= 630

What is 18 as a percentage of 50?
= 18/50
= (18/100) x 2 ---> method: division by 50 (explained above)

= 0.18 x 2
= 0.36
= 36%

What is 132 as a percentage of 200?
= 132/200
= (132/2)/100
= [100/2 + 32/2]/100 ---> solution by "double division"
= (50 + 16)/100
= 66/100
= 0.66
= 66%

What is 270 as a percentage of 300?
= 270/300
= [(270/3)/100] ---> "double division" (using the factors of 300)
= 90/100
= 90%

What is 17 as percentage of 500?
= 17/500
= (17/50)/10
= (17/100) x 2/10 ---> solution using the procedure: division by 50
= (0.17 x 2)/10
= 0.34/10
= 0.034
= 3.4 %

percentages close to 100:

SECRETS & TRICKS OF MATHEMATICS

95% of 700
= (100% - 5%) x 700
= (100% x 700) - (5% x 700)
= 700 - (10% x 700/2) -------> 5% is 10%/2
= 700 - 70/2
= 700 - 35
= 665

percentages less than 10 percent:
3% of 70
= (3/100) x 70
= (70/100) x 3 ---> divide by 100 then multiply the percent value
= 0.7 x 3
= 2.1

DECIMALS:

To convert or express percentages as decimals, divide by 100 or simply just move the decimal point by two places to the left of the given number, thus:

1% = 1/100 = 0.01
2% = 2/100 = 0.02 = 1/50
3% = 3/100 = 0.03
4% = 4/100 = 0.04 = 1/25
5% = 5/100 = 0.05 = 1/20
6.25% = 6.25/100 = 0.0625 = 1/16
7% = 7/100 = 0.07
7.5% = 7.5/100 = 0.075
10% = 10/100 = 0.1 = 1/10
12.5% = 12.5/100 = 0.125 = 1/8
20% = 0.2 = 1/5
21% = 0.21
25% = 0.25 = 1/4
30% = 0.3 = 3/10
33.33% = 33.33/100 = 0.3333 = 1/3
37.5% = 0.375 = 3/8
40% = 0.4 = 2/5
50% = 0.5 = 1/2
60% = 0.6 = 3/5
62.5% = 0.625 = 5/8
66.66% = 66.66/100 = 2/3
75% = 0.75 = 3/4
80% = 0.8 = 4/5
87.5% = 0.875 = 7/8

100% = 1
125% = 1.25 = 1 1/4
150% = 1.5 = 1 1/2
200% = 2

SECRETS & TRICKS OF MATHEMATICS

FRACTIONS:

What is three quarters of 80?
= 3/4 x 80
= (80/4) x 3 ---> divide by 4 then multiply by 3
= 20 x 3
= 60

How many quarters in two and a half?
2.5/.25
= 10 ---> there are 10 quarters in $2.50

Improper fractions:

3/2 = 1 1/2 = 1.5 = 150%

4/3 = 1 1/3 = 1.3333 = 133.33% ---> useful number for volume of sphere, etc.

9/5 = 1 4/5 = 1.8 = 180% ---> conversion factor for Celsius/Fahrenheit temperatures

$V = 4/3 \text{ pi} * r^3$

where:

V = volume of sphere
r = radius of sphere

$$F = (1.8\,C) + 32$$

where:

F = temperature in Fahrenheit
C = temperature in Celsius

http://useful-mathematics.blogspot.co.uk/2012/10/shortcuts-in-multiplication-division.html

Mental Math Shortcuts

Here's a collection of time-saving math shortcuts, great for back-of-the-envelope estimates.

Time and Distance

60 mph = 1 mile per minute

- Going 60 mph and the exit is in 10 miles? That's 10 minutes.
- Been driving a half hour? That's about 30 miles at highway speeds.

Feet Per Second = MPH * 1.5 MPH = Feet Per Second * 2/3 (derivation)

- 60 mph is about 90 feet per second (88 exactly), so just multiply by 1.5. Or, just add half to itself (60 + 30 = 90).
- Going 100 mph? That's 150 fps.
- Going 10 fps? That's about 7 mph (10 * 2/3 is 6.666). Or, just take away 1/3 (10 – 3 = 7).

speed of light = 1 foot per nanosecond (derivation)

- The US is about 3000 miles long. There's about 5000 feet/mile, so that's about 3000 × 5000 or 15 million feet. 15 million feet takes 15 million nanoseconds, or 15/1000, or 15 milliseconds. That's the minimum time for a signal to go across the country.
- Inside a microchip, if you have a clock cycle every nanosecond (1 GHz), your signal can only travel 1 foot at most (or less, depending on the material). Even light takes 30ns to cross a 30 foot room.

1 year = 250 work days = 2000 work hours (derivation)

- Project takes 1000 man hours? That's 6 months for 1 person.
- Daily commute of 1/2 hour? That's .5 * 250 = 125 hours in the car each year.

SECRETS & TRICKS OF MATHEMATICS

Money and Finance

$1/hour = $2000/year (derivation)

- Earn $25/hour? That's about 50k/year.
- Make 200k/year? That's about $100/hour. This assumes a 40-hour work week.

$20/week = $1000/year (derivation)

- Spend $20/week at Starbucks? That's a cool grand a year.

Rule of 72: Years To Double = 72/Interest Rate (derivation)

- Have an investment growing at 10% interest? It will double in 7.2 years.
- Want your investment to double in 5 years? You need 72/5 or about 15% interest.
- Growing at 2% a week? You'll double in 72/2 or 36 weeks. You can use this rule for any duration of time, not just years.
- Inflation at 4%? It will halve your money in 72/4 or 18 years.

Mental Arithmetic

Numbers

10,000 = hundred hundred million = thousand thousand billion = thousand million trillion = million million

- 1% of 10k is 100. The Dow is roughly 10k (it's about 12k now). So if the dow drops 100, it's about a 1% loss.

- What's 5k x 50k? That's 250 * thousand * thousand or 250 million.

Visualizing numbers (**read more**)

- 12 days = 1 million seconds
- 1 year = 31 million seconds (about pi * 10 million)
- 30 years = 1 billion seconds
- 30,000 years = 1 trillion seconds
- One "part per million" means an accuracy of 1 second every 12 days. One "part per trillion" means an accuracy of 1 second every 30,000 years.

Powers of 2

$2^6 = 64$ (the sixes match: six and sixty-four) 2^{10} ~ **thousand** (1 kb) 2^{20} ~ **million** (1 mb) 2^{30} ~ **billion** (1 gb)

- Sure, 2 to the tenth = 1024, but 1000 is good enough for government work. (Read on about **KB vs KiB**).
- Have 32-bit color? That's 2 + 30 bits = $2^2 * 2^{30} = 2^2$ billion = 4 billion (4gb exactly).
- Have a 16-bit number? That's 6 + 10 bits, or 2^6 thousand, or 64 thousand (64 kb).

a% of b = b% of a

- It's not immediately clear, but it's true: a% of b = .01 * a * b, which is the same as b% of a (.01 * b * a).
- What's 16% of 25? The same as 25% of 16: 4
- What's 43% of 200? Same as 200% of 43: 86.

SECRETS & TRICKS OF MATHEMATICS

Techniques for Adding the Numbers 1 to 100

There's a popular story that Gauss, mathematician extraordinaire, had a lazy teacher. The so-called educator wanted to keep the kids busy so he could take a nap; he asked the class to add the numbers 1 to 100.

Gauss approached with his answer: 5050. So soon? The teacher suspected a cheat, but no. Manual addition was for suckers, and Gauss found a formula to sidestep the problem:

$$\text{Sum from 1 to n} = \frac{n(n+1)}{2}$$

$$\text{Sum from 1 to 100} = \frac{100(100+1)}{2} = (50)(101) = 5050$$

Let's share a few explanations of this result and really understand it intuitively. For these examples we'll add 1 to 10, and then see how it applies for 1 to 100 (or 1 to any number).

Technique 1: Pair Numbers

Pairing numbers is a common approach to this problem. Instead of writing all the numbers in a single column, let's wrap the numbers around, like this:

SECRETS & TRICKS OF MATHEMATICS

$$\begin{array}{ccccc} 1 & 2 & 3 & 4 & 5 \\ 10 & 9 & 8 & 7 & 6 \end{array}$$

An interesting pattern emerges: **the sum of each column is 11**. As the top row increases, the bottom row decreases, so the sum stays the same.

Because 1 is paired with 10 (our n), we can say that each column has (n+1). And how many pairs do we have? Well, we have 2 equal rows, we must have n/2 pairs.

$$\text{Number of pairs * Sum of each pair} = (\frac{n}{2})(n+1) = \frac{n(n+1)}{2}$$

which is the formula above.

Wait — what about an odd number of items?

Ah, I'm glad you brought it up. What if we are adding up the numbers 1 to 9? We don't have an even number of items to pair up. Many explanations will just give the explanation above and leave it at that. I won't.

Let's add the numbers 1 to 9, but instead of starting from 1, let's count from 0 instead:

$$\begin{array}{ccccc} 0 & 1 & 2 & 3 & 4 \\ 9 & 8 & 7 & 6 & 5 \end{array}$$

By counting from 0, we get an "extra item" (10 in total) so we can have an even number of rows. However, our formula will look a bit different.

ns of n (not n+1, like before), since 0 and 9 are grouped. And instead of having exactly n items in 2 rows (for n/2 pairs total), we have n + 1 items in 2 rows (for (n + 1)/2 pairs total). If you plug these numbers in you get:

$$\text{Number of pairs} * \text{Sum of each pair} = \left(\frac{n+1}{2}\right)(n) = \frac{n(n+1)}{2}$$

which is the same formula as before. It always bugged me that the same formula worked for both odd and even numbers – won't you get a fraction? Yep, you get the same formula, but for different reasons.

Technique 2: Use Two Rows

The above method works, but you handle odd and even numbers differently. Isn't there a better way? Yes.

Instead of looping the numbers around, let's write them in two rows:

1	2	3	4	5	6	7	8	9	10
10	9	8	7	6	5	4	3	2	1

Notice that we have 10 pairs, and each pair adds up to 10+1.

The total of all the numbers above is

$$\text{Total} = \text{pairs} * \text{size of each pair} = n(n+1)$$

But we only want the sum of one row, not both. So we divide the formula above by 2 and get:

$$\frac{n(n+1)}{2}$$

Now this is cool (as cool as rows of numbers can be). It works for an odd or even number of items the same!

Technique 3: Make a Rectangle

I recently stumbled upon another explanation, a fresh approach to the old pairing explanation. Different explanations work better for different people, and I tend to like this one better.

Instead of writing out numbers, pretend we have beans. We want to add 1 bean to 2 beans to 3 beans… all the way up to 5 beans.

x
x x
x xx
x xxx
x xxxx

Sure, we could go to 10 or 100 beans, but with 5 you get the idea. How do we count the number of beans in our pyramid?

Well, the sum is clearly 1 + 2 + 3 + 4 + 5. But let's look at it a different way. Let's say we mirror our pyramid (I'll use "o" for the mirrored beans), and then topple it over:

```
x o x o oooo
x x o o x x o ooo
x xx o oo  =>  x xx o oo
x xxx o ooo x xxx o o
x xxxx o oooo x xxxx o
```

Cool, huh? In case you're wondering whether it "really" lines up, it does. Take a look at the bottom row of the regular pyramid, with 5'x (and 1 o). The next row of the pyramid has 1 less x (4 total) and 1 more o (2 total) to fill the gap. Just like the pairing, one side is increasing, and the other is decreasing.

Now for the explanation: How many beans do we have total? Well, that's just the area of the rectangle.

We have n rows (we didn't change the number of rows in the pyramid), and our collection is (n + 1) units wide, since 1 "o" is paired up with all the "x"s.

$$Area = height \cdot width = n(n+1)$$

Notice that this time, we don't care about n being odd or even – the total area formula works out just fine. If n is odd, we'll have an even number of items (n+1) in each row.

But of course, we don't want the total area (the number of x's and o's), we just want the number of x's. Since we doubled the x's to get the o's, the x's by themselves are just half of the total area:

$$\text{Number of x\^as} = \frac{Area}{2} = \frac{n(n+1)}{2}$$

And we're back to our original formula. Again, the number of x's in the pyramid = 1 + 2 + 3 + 4 + 5, or the sum from 1 to n.

Technique 4: Average it out

We all know that

average = sum / number of items

which we can rewrite to

sum = average * number of items

So let's figure out the sum. If we have 100 numbers (1…100), then we clearly have 100 items. That was easy.

To get the average, notice that the numbers are all equally distributed. For every big number, there's a small number on the other end. Let's look at a small set:

1 2 3

The average is 2. 2 is already in the middle, and 1 and 3 "cancel out" so their average is 2.

For an even number of items

1 2 3 4

the average is between 2 and 3 – it's 2.5. Even though we have a fractional average, this is ok — since we have an **even** number of items, when we multiply the average by the count that ugly fraction will disappear.

Notice in both cases, 1 is on one side of the average and N is equally far away on the other. So, we can say the average of the entire set is actually just the average of 1 and n: (1 + n)/2.

Putting this into our formula

$$\text{sum} = \text{average} * \text{count} = \frac{(1+n)}{2} \cdot n = \frac{n(n+1)}{2}$$

And voila! We have a fourth way of thinking about our formula.

So why is this useful?

Three reasons:

1) Adding up numbers quickly can be useful for estimation. Notice that the formula expands to this:

$$\frac{n(n+1)}{2} = \frac{n^2}{2} + \frac{n}{2}$$

Let's say you want to add the numbers from 1 to 1000: suppose you get 1 additional visitor to your site each day – how many total visitors will you have after 1000 days? Since thousand squared = 1 million, we get million / 2 + 1000/2 = 500,500.

2) This concept of adding numbers 1 to N shows up in other places, like figuring out the probability for the birthday paradox. Having a firm grasp of this formula will help your understanding in many areas.

3) Most importantly, this example shows there are many ways to understand a formula. Maybe you like the pairing method, maybe you prefer the rectangle technique, or maybe there's another explanation that works for you. **Don't give up** when you don't understand — try to find another explanation that works. Happy math.

By the way, there are more details about the history of this story and the technique Gauss may have used.

Variations

Instead of 1 to n, how about 5 to n?

Start with the regular formula $(1 + 2 + 3 + \ldots + n = n * (n + 1) / 2)$ and subtract off the part you don't want $(1 + 2 + 3 + 4 = 4 * (4 + 1) / 2 = 10)$.

Sum for $5 + 6 + 7 + 8 + \ldots n = [n * (n + 1) / 2] - 10$

And for any starting number a:

Sum from a to n $= [n * (n + 1) / 2] - [(a - 1) * a / 2]$

We want to get rid of every number from 1 up to $a - 1$.

How about even numbers, like 2 + 4 + 6 + 8 + … + n?

Just double the regular formula. To add evens from 2 to 50, find 1 + 2 + 3 + 4 … + 25 and double it:

Sum of 2 + 4 + 6 + … + n = 2 * (1 + 2 + 3 + … + n/2) = 2 * n/2 * (n/2 + 1) / 2 = n/2 * (n/2 + 1)

So, to get the evens from 2 to 50 you'd do 25 * (25 + 1) = 650

How about odd numbers, like 1 + 3 + 5 + 7 + … + n?

That's the same as the even formula, except each number is 1 less than its counterpart (we have 1 instead of 2, 3 instead of 4, and so on). We get the next biggest even number (n + 1) and take off the extra (n + 1)/2 "-1" items:

Sum of 1 + 3 + 5 + 7 + … + n = [(n + 1)/2 * ((n + 1)/2 + 1)] − [(n + 1) / 2]

To add 1 + 3 + 5 + … 13, get the next biggest even (n + 1 = 14) and do

[14/2 * (14/2 + 1)] − 7 = 7 * 8 − 7 = 56 − 7 = 49

Combinations: evens and offset

Let's say you want the evens from 50 + 52 + 54 + 56 + … 100. Find all the evens

2 + 4 + 6 + … + 100 = 50 * 51

and subtract off the ones you don't want

$2 + 4 + 6 + \ldots 48 = 24 * 25$

So, the sum from $50 + 52 + \ldots 100 = (50 * 51) - (24 * 25) = 1950$

Phew! Hope this helps.

Ruby nerds: you can check this using

(50..100).select {|x| x % 2 == 0 }.inject(:+)

1950

http://betterexplained.com/articles/mental-math-shortcuts/

April Fool Math

Ask all students/participants to write down a <u>three digit number</u>.

The first and third digits must differ by more than one.
Example : 264

Now *reverse the digits* to form a second number.
Example : 462

Subtract the smaller number from the larger one.
Example : 462 - 264 = 198

Now *reverse* the digits in the answer you got in step 3 and add it to that number.
Example : 891 + 198 = 1089

Multiply the number by *one million*.
Example : 1089 x 1000000 = 1089000000

Subtract 733,361,573
Example : 1089000000 - 733,361,573 = 355638427

SECRETS & TRICKS OF MATHEMATICS

Magic code

Under each of the digits in your answer, write the letter which corresponds to it using the following table:

0 - Y
1 - M
2 - P
3 - L
4 - R
5 - O
6 - F
7 - A
8 - I
9 - B

Example: 3 5 5 6 3 8 4 2 7
L O O F L I R P A

Now read your message backward.

http://www.pedagonet.com/videos/aprilfool.htm

Always 5

Find a calculator or a pencil and paper.
Ask your friend (or everyone in the room) to choose any number.
Example : 43

Add the next <u>highest</u> number to it
43 + **44** = **87**

Add **9**
87 + 9 = 96

Divide by **2**
96 / 2 = 48

Subtract your <u>original</u> number
48 - 43 = **5**

Everyone's answer will **always** be 5
Try another number.
More Always Five

SECRETS & TRICKS OF MATHEMATICS

Choose any number between 1 and 99
Example : 22
Multiply this number by **5**
22 x 5 = 110
Add **25** to your product
110 + 25 = 135
Divide this number by **5**
135 / 5 = 27
Subtract your <u>original</u> number
27 - 22 = 5
Everyone's answer will **always** be 5
One More Trick

Choose any number.
Double it.
Add 10 to it.
Now half the number.
Then subtract your first number.
The answer will be five.

http://www.pedagonet.com/maths/always.htm

Five Statistics Problems That Will Change The Way You See The World

WALTER HICKEY NOV. 13, 2012, 8:29 PM

Even a rudimentary look at probability can give new insights about how to interpret data.

Simple thought experiments can give new insight into the different ways a misunderstanding of statistics distorts the way we perceive the world.

We've selected five classic problems solved in unconventional ways that can help one get a new way to understand how data can be misleading and the story on the surface can take people in the wrong direction.

SECRETS & TRICKS OF MATHEMATICS

The Monty Hall Problem

Wikimedia Commons

Say you're on a game show where there are three doors. Behind two of the doors, there are goats. Behind one of the doors, there is a brand new car.

The host says that once you pick a door, he'll open one of the doors you didn't pick to reveal a goat. Then, you have the option of either staying with your door or switching to the last unopened door.

Do you switch or stay?

Answer: Switch

Flickr

This is actually based on a real game show, and the result has been the source of controversy for years.

Essentially, when you first made the selection, you had a one in three chance of correctly selecting the door that had a car behind it. Switching raised that probability to two in three that you'll select a car.

Said another way: A player whose strategy is to always switch will only lose when the door they initially selected has a car behind it. A contestant who selects either of the two doors with a goat behind it and then switches will always get the car.

Here's a final way to look at it, provided the contestant selected Door #1

Door 1	Door 2	Door 3	Result if Stay #1	Result if Switch
Car	Goat	Goat	Car	*Goat*
Goat	Car	Goat	Goat	**Car**
Goat	Goat	Car	Goat	**Car**

Source: *The Straight Dope*

SECRETS & TRICKS OF MATHEMATICS

The Birthday Paradox

Flickr/Cali4beach

You run an office that employs 23 people. What is the probability that two of your employees have the same birthday? For the purposes of the problem, ignore February 29.

Answer: 50%

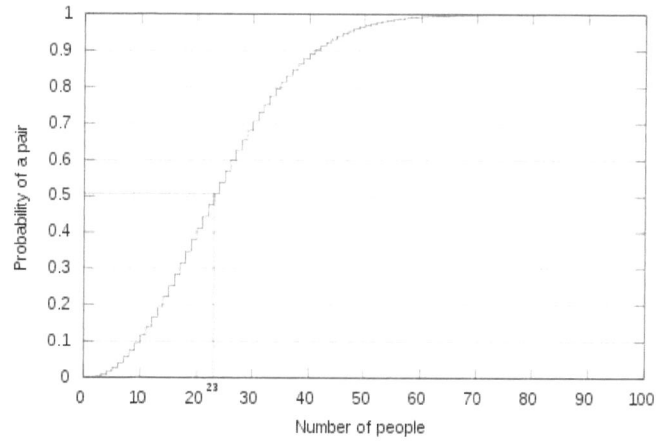

Rajkiran g / wikimedia

Once the population of an office hits 366 people, it's a certainty that two people in your office have the same birthday, since there are only 365 possible days of birth.

Still, assuming that each birth date (except February 29) is equally likely, it turns out that once your office has 57 people in it there is a 99% chance that two of them share a birthday. When there is 23 people, that probability is 50%.

Here's why. Instead of calculating the probability that two people share a birthday, instead calculate the converse, probability that two people don't share a birthday. Since these are mutually exclusive scenarios, first probability plus the second probability has to equal 1.

Here's how we figure this out, then.

Select two people in the office. The probability of the second person not sharing a birthday with the first is 364/365. The probability of the third person not sharing a birthday with the first or second is 363/365. Going through the office and multiplying these together, we see this:

365/365 x 364/365 x 363/365 x 362/365 x ... x 343/365 = 0.4927.

So, the probability that nobody in an office of 23 people share a birthday is 0.4927, or 49.3%. That means that the probability that two people in the office share a birthday is 1 — 0.4927 = 0.5073, or 50.7%.

Source: *Better Explained*

Gambler's Ruin

A gambler has a certain amount of money ("B") and is playing a game of chance with some win probability less than 1. Every time he wins, he raises his stake to a certain fraction, 1/N, of his bankroll, where N is a positive number. The gambler doesn't reduce his stake when he loses

Every time he wins, he'll raise his stake to $B/N, or his bankroll divided by N. When B= $1000 and N=4, for example, he'll gamble $250 each time going forward. Should he win, he'll raise it again. Should he lose, he'll keep his stake at $250.

If he keeps at it, what are his expected winnings?

Answer: He'll lose everything

SECRETS & TRICKS OF MATHEMATICS

Flickr - greengardenvienna

When it comes down to it, if our gambler bets 1/N of his bankroll each time and then maintains the amount as he loses, the gambler is N losing bets in a row away from bankruptcy.

Assuming that the player keeps on playing and there is some chance that the player can lose — we are gambling, after all — then the player remains N losing bets away from a broken bank each time.

If our gambler sounds like something of an idiot, know that this is actually a rather common betting strategy. Casinos also endorse it by ensuring that players are stocked with <u>mostly high denominational chips</u> as they go on winning streaks in order to encourage higher bets.

Even more, consider the ante in a game of poker, which is a similar system designed to accelerate a winner.

Source: <u>University of California San Diego</u>

Abraham Wald's Memo

Wikipedia

Abraham is tasked with reviewing damaged planes coming back from sorties over Germany in the Second World War. He has to review the damage of the planes to see which areas must be protected even more.

Abraham finds that the fuselage and fuel system of returned planes are much more likely to be damaged by bullets or flak than the engines. What should he recommend to his superiors?

Answer: Protect the parts that don't have damage

MathematischesForschungsinstitutOberwolfach

Abraham Wald, a member of the Statistical Research Group at the time, saw this problem and made an unconventional suggestion that saved countless lives.

Don't arm the places that sustained the most damage on planes that came back. By virtue of the fact that these planes came back, these parts of the planes can sustain damage.

If an essential part of the plane comes back consistently undamaged, like the engines in the previous example, that's probably because all the planes with shot-up engines don't make it back.

Wald's memos on this situation — in addition to being a remarkable historical statistical document — shed additional light of the statistics developed during the Second World War that would go on to found the field of Operations Research.

Source: Marc Mangel, FransiscoSamaniego

SECRETS & TRICKS OF MATHEMATICS

Simpson's Paradox

AP

A kidney study is looking at how well two different drug treatments (A and B) work on small and large kidney stones. Here is the success rate that was found:

- Small Stones, Treatment A: 93%, 81 out of 87 trials successful
- Small Stones, Treatment B: 87%, 234 out of 270 trials successful
- Large Stones, Treatment A: 73%, 192 out of 263 trials successful
- Large Stones, Treatment B: 69%, 55 out of 80 trials successful.
- All stones, Treatment A: 78%, 273 of 350 trials successful
- **All stones, Treatment B: 83%, 289 of 350 trials successful.**

Which is the better treatment, A or B?

Answer: Treatment A, once you focus on the subsets

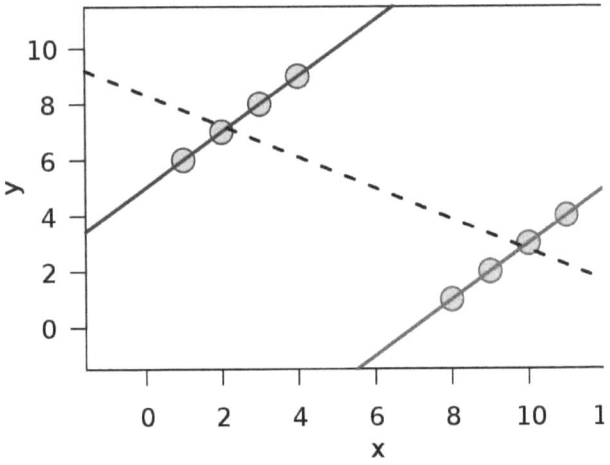

Schutz / Wikimedia

Even though Treatment A had higher success rates in both small and large stones, when the whole trial is viewed as a sample space Treatment B seemed more successful:

- **Small Stones, Treatment A: 93%, 81 out of 87 trials successful**
- Small Stones, Treatment B: 87%, 234 out of 270 trials successful
- **Large Stones, Treatment A: 73%, 192 out of 263 trials successful**
- Large Stones, Treatment B: 69%, 55 out of 80 trials successful.

- All stones, Treatment A: 78%, 273 of 350 trials successful
- **All stones, Treatment B: 83%, 289 of 350 trials successful.**

This is an excellent example of Simpson's Paradox, where correlation in separate groups doesn't necessarily translate to the whole sample set, causing ambiguity.

In short, just because there correlation in smaller groups hides the real story taking place in the largest of groups.

Source: Stephen Julious, Thanks to Kerry Champion

http://www.businessinsider.com/five-statistics-problems-that-will-change-the-way-you-see-the-world-2012-11?op=1

Steps to Doing Well in Math

Improve Your Math Marks

//////////By Deb Russell

Here are some quick steps to help you get better at doing mathematics. Regardless of age, the tips here will help you learn and understand math concepts from primary school right on through to university math. Everyone can do math, be positive and follow the steps here and you'll be on your way to seeing success in math.

Understanding Versus Memorizing

Getty Images

All too often, we will try to memorize a procedure or sequence of steps instead of looking to understand why certain steps are required in a procedure. Always, always strive for understanding the why and not just the how. Take the algorithm for <u>long division</u> Typically, we say, "how many times does 3 go into 7" when the question is 73 divided by 3. After all, that 7 represents 70 or 7 tens. The understanding in this question really has little to do with how many times 3 goes into 7 but rather **how many** are in the group of three when you share the 73 into 3 groups. 3 going into 7 is merely a short cut. Putting 73 into 3 groups means understanding. The long division algorithm rarely makes sense unless the concrete method is fully understood.

Getty Images

Unlike some subjects, math is something that won't let you be a passive learner. Math is the subject that will often put you out of the comfort zone, don't worry as this is normal and part of the learning process. Try to make connections in math, many of the concepts in math are related and connected. The more connections you can make, the greater the understanding will be. Math concepts flow through levels of difficulty, start from where you are and move forward to the more difficult levels only when understanding is in place. The internet has a wealth of interactive math sites that let you engage, be sure use them.

Practice, Practice, Practice

Getty Images

Do as many problems as is required to ensure you understand the concept. Some of us require more practice and some of us require less practice. You will want to practice a concept until it makes sense and until you are fluent at finding solutions to various problems within the concept readily. Strive for those 'A Ha!' moments. When you can get 7 varied questions in a row right, you're probably to the point of understanding. Even more so if you re-visit the questions a few months later and are still capable of solving them. This too is key to understanding. Be sure to check out the worksheet section for lots of practice examples.

Additional Exercises

Getty Images

This is similar to practice. Think of math the way one thinks about a musical instrument. Most of us don't just sit down and play an instrument. We take lessons, practice, practice some more and although we move on, we still take time to review. Go beyond what is asked for. Your instructor tells you to do questions 1-20, even numbers only. Well, that may work for some, but others may need to do each of the questions to reach the point of fluency with the concept. Doing the extra practice questions only helps you to grasp the concept more readily. And, as always, be sure to re-visit a few months later, do some practice questions to ensure that you still have a grasp of it.

Buddy Up!

Getty Images

Some people like to work alone. However, when it comes to solving problems, it often helps to have a work buddy. You know the saying: two heads are better than one. Sometimes a work buddy can help clarify a concept for you by looking at it in a different way. Organize a study group or work in pairs or triads! In real life we often work through problems with others. Math is no different. A work buddy also provides you with the opportunity to discuss how you solved the math problem. And as you'll see in this list of tips, conversing about math leads too permanent understanding and you know that understanding is key.

Explain and Question

Getty Images

Try to explain to somebody else how you solve math concepts. Teach a friend. Or, keep a journal. It's often important to state either in writing or orally how you solved your math problems/exercises. Question problems, ask yourself, What would happen if.......I solved it this way because.....

Remember William Glasser's findings: 10% of what we READ

20% of what we HEAR
30% of what we SEE
50% of what we SEE and HEAR
70% of what is DISCUSSED with OTHERS
80% of what is EXPERIENCED PERSONALLY
95% of what we TEACH TO SOMEONE ELSE

William Glasser

Phone a Friend……or Tutor!

Getty Images

Seek help when it's appropriate. Don't let yourself get stuck and frustrated. Seek extra clarification when needed, be your own advocate! Whether you have a friend or need to hire a tutor, recognize the point at which you need help - then get it! Most of us need help some of the time, if you let it go too long, you'll discover that the math will only become more frustrating for you.

http://math.about.com/od/booksresourcesdvds/tp/Improve-Math.htm

Math Magic/Tricks

http://easycalculation.com/funny/tricks/trick1.php

Trick 1: Number below 10

Step1:
Think of a number below 10.
Step2:
Double the number you have thought.
Step3:
Add 6 with the getting result.
Step4:
Half the answer, that is divide it by 2.
Step5:
Take away the number you have thought from the answer, that is, subtract the answer from the number you have thought.

Answer: 3

Trick 2: Any Number

Step1:
Think of any number.
Step2:
Subtract the number you have thought with 1.
Step3:

Multiply the result with 3.
Step4:
Add 12 with the result.
Step5:
Divide the answer by 3.
Step6:
Add 5 with the answer.
Step7:
Take away the number you have thought from the answer, that is, subtract the answer from the number you have thought.

Answer: 8

Trick 3: Any Number
Step1:
Think of any number.
Step2:
Multiply the number you have thought with 3.
Step3:
Add 45 with the result.
Step4:
Double the result.
Step5:
Divide the answer by 6.
Step6:
Take away the number you have thought from the answer, that is, subtract the answer from the number you have thought.

Answer: 15

Trick 4: Same 3 Digit Number

Step1:
Think of any 3 digit number, but each of the digits must be the same as. Ex: 333, 666.
Step2:
Add up the digits.
Step3:
Divide the 3 digit number with the digits added up.

Answer: 37

Trick 5: 2 Single Digit Numbers

Step1:
Think of 2 single digit numbers.
Step2:
Take any one of the number among them and double it.
Step3:
Add 5 with the result.
Step4:
Multiply the result with 5.
Step5:
Add the second number to the answer.
Step6:
Subtract the answer with 4.
Step7:
Subtract the answer again with 21.

Answer: 2 Single Digit Numbers.

SECRETS & TRICKS OF MATHEMATICS

Trick 6: 1, 2, 4, 5, 7, 8

Step1:
Choose a number from 1 to 6.
Step2:
Multiply the number with 9.
Step3:
Multiply the result with 111.
Step4:
Multiply the result by 1001.
Step5:
Divide the answer by 7.

Answer: All the above numbers will be present.

Trick 7: 1089

Step1:
Think of a 3 digit number.
Step2:
Arrange the number in descending order.
Step3:
Reverse the number and subtract it with the result.
Step4:
Remember it and reverse the answer mentally.
Step5:
Add it with the result, you have got.

Answer: 1089

Trick 8: x7x11x13

Step1:
Think of a 3 digit number.
Step2:
Multiply it with x7x11x13.
Ex: Number: 456, Answer: 456456

Trick 9: x3x7x13x37

Step1:
Think of a 2 digit number.
Step2:
Multiply it with x3x7x13x37.
Ex: Number: 45, Answer: 454545
Trick 10: 9091
Step1:
Think of a 5 digit number.
Step2:
Multiply it with 11.
Step3:
Multiply it with 9091.

Ex: Number: 12345, Answer: 1234512345

Is it Magic or Is it Maths?

Guess how much money people have in their pockets! Without giving you any information, ask a friend to count the value of some coins and write the amount on a piece of paper. Then ask your friend to:

- Double the amount.
- Add the first odd prime number to the new total.
- Multiply the result by 1/4 of 20.
- Subtract the lowest common multiple of 2 and 3.

For the grand finale, you ask for the final answer. Take off the last digit and you will be able to work out how much the coins are worth!

 Amaze your audience by working out not only their age but also what size shoe they wear! Wow them even more by telling them how the maths works.

Give them the following directions but tell them not to show you any calculations: Write down your age.

- Multiply it by 1/5 of 100.
- Add on today's date (e.g. 2 if it's the 2nd of the month).

SECRETS & TRICKS OF MATHEMATICS

- Multiply by 20% of 25.
- Now add on your shoe size (if it's a half size round to a whole number).
- Finally subtract 5 times today's date.
- Show me you final answer!

Look at the answer, the hundreds are the age and the remaining digits are the shoe size. If for instance somebody shows you 1105, there are 11 hundreds - the age, and the remaining digits 05 (or 5) show the shoe size.

Now, how on earth does that work?

http://nrich.maths.org/1051

The Calendar Trick

This trick makes you look like you've got the most awesome mental power. All you need is a calendar which has the dates lined up under the days of the week, so the numbers are arranged something like this:

.	.	1	2	3	4	5
6	7	8	9	10	11	12
13	14	15	16	17	18	19
20	21	22	23	24	25	26
27	28	29	30	31		

The "9 number" trick

- Ask a friend to draw a 3x3 box around ANY nine of the numbers on the calendar.
- Almost immediately you can say what the nine numbers all add up to!
- See how long it takes your friend to check the answer on a calculator. (Yawn...snore...zzzz...)

For instance, let's say the red numbers are chosen.

		1	2	3	4	5
6	7	8	9	10	11	12
13	14	15	16	17	18	19
20	21	22	23	24	25	26
27	28	29	30	31		

They add up to 198.

THE SECRET: *All you do is look at the number in the middle and multiply it by 9.*

In this case the middle number is 22 and 22 x 9 = 198.

HANDY TIP: To multiply by 9 quickly, just multiply by 10 then subtract your number.

So to get 22 x 9 you multiply 22 x 10 =220 (easy!) and then subtract 22. With a bit of practise you can do this quickly in your head.

The "20 number" trick

- Ask a friend to draw a 5x4 box around ANY twenty of the numbers on the calendar.
- Almost immediately you can say what all twenty numbers add up to!

For instance, let's say the green numbers are chosen.

	1	2	3	4	5	6
7	**8**	9	10	11	12	13
14	**15**	16	17	18	19	20
21	**22**	23	24	25	26	27
28	**29**	**30**	**31**			

Thanks to Louise Lennartsson who spotted a mistake in the calendar grid on the left (which we've corrected). We estimate that over 50,000 people had already viewed this page before she saw it!.

They add up to 290.

THE SECRET: *Add together the smallest and the largest numbers in the group. Multiply the answer by 10.*

In this case the smallest number is 2 and the largest is 27, so 2 + 27 = 29. Then 29 x 10 = 290.

With the "20 number trick" if the calendar month is February occasionally it will not be possible to put a box round 20 numbers.

SECRETS & TRICKS OF MATHEMATICS

Here's this months calendar for you to practise on!

```
              May 2014
    Su   M   Tu   W   Th   F   Sa
                      1    2   3
    4    5   6    7   8    9   10
    11   12  13   14  15   16  17
    18   19  20   21  22   23  24
    25   26  27   28  29   30  31
         |   |    |   |    |
```

And if you don't want to use a calendar....

These tricks work with ANY grid of numbered boxes, just as long as the numbers are continuous. Try them both on this grid:

```
13  14  15  16  17  18  19  20  21  22  23
24  25  26  27  28  29  30  31  32  33  34
35  36  37  38  39  40  41  42  43  44  45
46  47  48  49  50  51  52  53  54  55  56
57  58  59  60  61  62  63  64  65  66  67
68  69  70  71  72  73  74  75  76  77  78
```

http://www.murderousmaths.co.uk/games/calendar.htm

THE SECRET WORLD OF CODES AND CODE BREAKING

Stage: 2, 3 and 4
Article by NRICH team

When you think of spies and secret agents, you might think of lots of things; nifty gadgets, foreign travel, dangerous missiles, fast cars and being shaken but not stirred. You probably wouldn't think of mathematics. But you should.

Cracking codes and unravelling the true meaning of secret messages involves loads of maths, from simple addition and subtraction, to data handling and logical thinking. In fact, some of the most famous code breakers in history have been mathematicians who have been able to use quite simple maths to uncovered plots, identify traitors and influence battles.

The Roman Geezer

Let me give you an example. Nearly 2000 years ago, Julius Caesar was busy taking over the world, invading countries to increase the size of the Roman Empire. He needed a way of communicating his battle plans and tactics to everyone on his side without the enemy finding out. So Caesar would write messages to his generals in code. Instead of writing the letter 'A', he would write the letter that comes three places further on in the alphabet, the letter 'D'. Instead of a 'B', he would write an 'E', instead of a 'C', he would write an 'F' and so on. When he got to the end of the alphabet, however, he would have to go right back to the beginning, so instead of an 'X', he would write an 'A', instead of a 'Y', he'd write a 'B' and instead of 'Z', he'd write a 'C'.

Complete the table to find out how Caesar would encode the following message:

Caesar's message	A	T	T	A	C	K	A	T	D	A	W	N
	B	U										
	C	V										
Coded message	D											

When Caesar's generals came to decipher the messages, they knew that all they had to do was go back three places in the alphabet. Have a go at trying to work out these messages which could have been sent by Caesar or his generals:

hqhpbdssurdfklqj
wkluwbghdg
uhwuhdwwriruhvw

Easy as 1, 2, 3

This all seems very clever, but so far it's all been letters and no numbers. So where's the maths? The maths comes if you think of the letters as numbers from 0 to 25 with A being 0, B being 1, C being 2 etc. Then encoding, shifting the alphabet forward three places, is the same as adding three to your starting number:

A	B	C	D	E	F	G	H	I	J	K	L	M	N
0	1	2	3	4	5	6	7	8	9	10	11	12	13

O	P	Q	R	S	T	U	V	W	X	Y	Z
14	15	16	17	18	19	20	21	22	23	24	25

For example, encoding the letter 'A' is 0+3=3, which is a 'D'.

Coding 'I' is: 8+3=11, which is 'L'.

However, you do have to be careful when you get to the end of the alphabet, because there is no letter number 26, so you have to go back to number 0. In maths we call this 'MOD 26', instead of writing 26, we go back to 0.

Have a go at coding your name by adding 3 to every letter. Then have a go at coding your name by shifting the alphabet forward by more places by adding greater numbers eg adding 5, then adding 10. Then have a go at decoding. If your letters are numbers and encoding is addition, then decoding is subtraction, so if you've coded a message by adding 5, you will have to decode the message by subtracting 5.

Treason!

If you've got the hang of coding messages by shifting the alphabet forward, then you might have realised that it is actually pretty simple to crack this type of code. It can easily be done just by trial and error. An enemy code breaker would only have to try out 25 different possible shifts before they were able to read your messages, which means that your messages wouldn't be secret for very long.

So, what about coding messages another way? Instead of writing a letter, we could write a symbol, or draw a picture. Instead of an 'A' we could write *, instead of a 'B' write + etc. For a long time, people thought this type of code would be really hard to crack. It would take the enemy far too long to figure out what letter of the alphabet each symbol stood for just by trying all the possible combinations of letters and symbols. There are 400 million billion billion possible combinations!

This type of code was used by Mary Queen of Scots when she was plotting against Elizabeth the First. Mary wanted to kill Elizabeth so that she herself could become Queen of England and was sending coded messages of this sort to her co-conspirator Anthony Babington. Unfortunately for Mary, there is a very simple way of cracking this code that doesn't involve trial and error, but which does involve, surprise, surprise, maths.

SECRETS & TRICKS OF MATHEMATICS

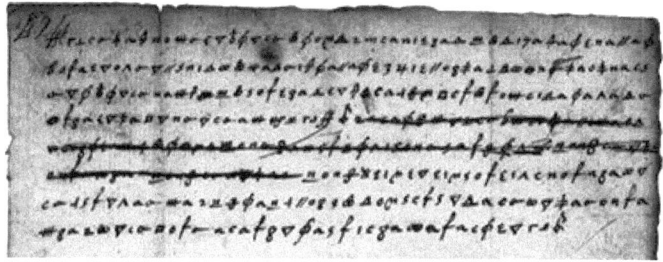

Letter sent by Mary Queen of Scots to her co-conspirator Anthony Babington. Every symbol stands for a letter of the alphabet.

Letters in a language are pretty unusual because some get used more often than other letters. An easy experiment you can do to test this out is to get everyone in your class to raise their hand if they have the letter 'E' in their name. Then get all those with a 'Z' to raise their hand, then a 'Q', then an 'A'. You will probably find that 'E' and 'A' are more common than 'Z' and 'Q'. The graph below shows the average frequency of letters in English. To compile the information, people looked through thousands and thousands of books, magazines and newspapers, and counted the number of times each letter came up.

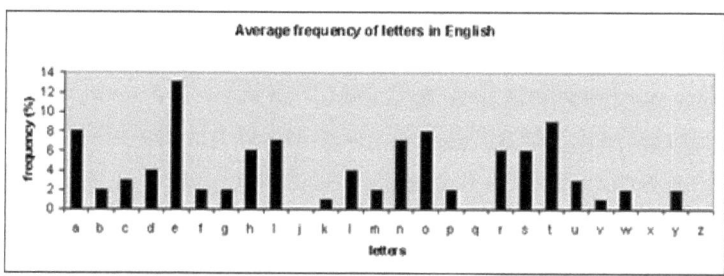

In English, E is the most commonly used letter. In any piece of writing, we use E about 13% of the time on average. 'T' is the second most common letter and 'A' is the third most commonly used letter.

And it's this information that can help you to crack codes. All Elizabeth the First's Spy-Master had to do to crack Mary's code, was to look through the coded message and count the number of times each symbol came up. The symbol that came up the most would probably stand for the letter 'E'. Look at our <u>Ancient Runes</u> problem for another code that could be deciphered by counting how often each symbol appears.

When you crack codes like this, by looking for the most common letter, it's called 'frequency analysis', and it was this clever method of cracking codes that resulted in Mary having her head cut off. CHOP!

Test your talents

Cracking these coded messages doesn't just involve looking for the most common symbol, you can also look for symbols that are all out on their own in the message ie one letter words. There are only two one-letter words in English, 'A' and 'I', so a lone symbol would have to stand for an 'A' or 'I'. Another thing you can look out for are common words. The most common three letter words in English are 'the' and 'and', so if you see a group of three symbols that comes up quite a lot, they could stand for 'the' or 'and'.

SECRETS & TRICKS OF MATHEMATICS

If you would like to test out these code breaking tips and your new code breaking talents, have a look at <u>Simon Singh's Black Chamber</u>. It has Caesar shift and frequency analysis puzzles for you to break, and other codes that you can try to unravel.

For more information about other secret codes that have been used throughout history, check out <u>Simon Singh's web site</u>. It's packed full of information about all sorts of codes, including the famous story Enigma, the code machine used by the Germans during WWII. The Germans thought their code was invincible, but incredibly, British mathematicians managed to break the code and read all the messages sent by the Germans during the war. Historians think that having this inside information shortened the war by two whole years.

WARNING

After reading this, you might fancy making up some codes of your own, and writing you own secret messages. BE WARNED. Other people have also read this article and they too will be top mathematical codebreakers. Spies are everywhere, so be careful - who's reading your messages?

Claire Ellis, the author of this article, was director of the Enigma Project, which takes codes and code breaking, and a genuine WW2 Enigma machine, into the classroom. For more information contact the new director, Claire Greer, via the <u>Enigma Schools' Project web site</u>.

<u>http://nrich.maths.org/2197</u>

The Secret to Success in Mathematics

There are so many people who struggle with mathematics and yet some people seem to have no trouble whatsoever. Is there something different in their DNA or some super advanced part of their brains, or do they simply know some secret trick that helps them in their understanding?

The answer is a little of the second, but mostly the last. Scientific research suggests that those who are good at mathematics do have some areas of their brain more active during mathematical activities, however, this increased brain activity could be a result of increased mathematical training. Much like how an athlete's muscles improve with training, regular mathematical exercises help to improve the performance of those brain areas associated with mathematics. Again, like an athlete in training, there needs to be some underlying skills and a good coach. All of us possess the minimum underlying skills, but not all of us are fortunate enough to have a good coach at the critical time when we are ready to start serious mathematical training.

That's not to say that you must have a brilliant mathematics teacher in order to succeed, but rather that when you are ready to start learning about mathematics beyond simple addition and subtraction, you need someone who can help you identify that mathematics is not a series of boring repetitive exercises from a textbook. Nor is mathematics made up of a set of discrete topics that package neatly into textbook chapters. In short, you need someone to tell you the secret to success in mathematics.

The secret to success in mathematics is understanding and accepting the following:

1. **Mathematics is a language**
   Mathematics is not a set of confusing hieroglyphs designed by some evil conspiracy to torture school children and university students, but rather a language built on rules. In the same way we learn any language, we must learn the alphabet and the rules used if we are to have any success in expressing our ideas or understanding others using this language. The primary rule of the language of mathematics is the "Order of Operations".
2. **Mathematics is based in logic**

Mathematical rules are firmly based in logic. For example, if we accept that 1<2 and 2<3, then it follows that 1<3. Whilst this is a simple example, complex mathematical rules are built by joining together many more simple rules. When the logic is violated, the result can be a set of mathematics that seems to prove the ridiculous, such as that 1 = 2 (see <u>post</u>). If you are careful about how you apply the rules and ensure you don't violate the logic behind them, you will be unlikely to go wrong.

3. **Mathematics is interconnected**

    Like many things in life, mathematical ideas are often connected to more than one other idea. Techniques, such as those that apply to linear functions, often reappear in other areas of mathematics, for example, in the statistical technique of Simple Linear Regression. As such, it is dangerous to treat mathematics as a set of discrete skills, learned for one chapter of a text and forgotten shortly after the topic test. Instead, one should consider each newly acquired skill or technique as part of an arsenal or toolkit to be drawn upon for future problems.

4. **Mathematics is everywhere**

    Much of mathematics taught in schools and universities suffers from an inability to answer the highly intelligent question, "Where will I use this in real life?". Unfortunately, it is often difficult to point to a real world situation and make a plausible justification for the direct use of much of the mathematics taught in high school. Pythagora's theorem ($a^2 + b^2 = c^2$) is a classic case. The ancients knew that if you took a rope, placed knots at equal distances and then used this to construct a triangle that had sides of 3 knots, 4 knots and 5 knots, then the triangle contained a right angle (very useful for constructing buildings that won't easily fall down). Despite this, it is unlikely that many modern students will need to build a pyramid for their pharaoh, so why should they learn it? Well, Pythagora's theorem is about more than right-angled triangles, it also tells us about the relationship between many numbers and is a theorem that is the basis of much of the mathematics used in modern computing, communications technology and cryptography.

5. **Mathematicians are lazy**

   Like many people, mathematicians don't want to do more work than they have to. As such, many of the techniques employed in mathematics are about reducing a problem down to a set of previously solved sub-problems. In other words, we mathematicians want to re-use our previous work (or someone else's work) wherever possible. What this often means is that students are often presented with almost identical problems as they learn different techniques. Unfortunately, this can lead to the perception that mathematics is nothing more than a set of boring repetitive exercises from a textbook, which is completely not true. Instead, perhaps it would be better to think that mathematics is very eco-friendly, recycling previous problems for reuse in more complex learning situations.

6. **Mathematicians like simplicity**

   The axiom of "The simplest solution is often the best" is truly at home in mathematics. As such, we prefer solutions that are given in the simplest form, whether that is as a surd, a fraction, a decimal or a whole number. Wherever possible, you should try to express your solutions in the simplest, but most correct, manner. For example, one should write $\sqrt{2}$ rather than 1.414.

7. **Mathematicians like order**

   Some might suggest that mathematicians suffer from some kind of obsessive compulsive disorder, but the truth is that mathematics is based upon order, be that the counting order of numbers or the "Order of Operations". The same is true about how we like to write equations in descending order of powers, e.g. $y = x^2 - 3x + 5$ rather than $y = 5 + x^2 - 3x$.

8. **Statistics is a special kind of mathematics**

   Statistics uses all the techniques of mathematics, from simple algebra through to calculus and beyond, but it also uses a special set of skills called "Reasoning with Uncertainty". Statistics is a branch of mathematics devoted to the science of uncertainty (risk, odds, chance, probability, likelihood). Unfortunately, it has gained a fairly bad reputation, mostly because, like weather forecasts, it does not provide 100% guarantees about the results, only the methods used to obtain them. Instead, statistics provides insight into the possibilities, which, when combined with other facts, should lead one to a reasonable (but still possibly wrong) conclusion based upon the sample data. Mark Twain once quipped, "There are three kinds of lies: lies, damned lies and statistics", hinting at this possibility of an incorrect conclusion. Whilst statistics is the

"Science of Uncertainty", there is no uncertainty in the science of statistics. If you perform the same statistical technique many times on the same data, it will always lead to the same conclusion. The uncertainty comes from the data and whether or not it is an accurate and representative sample of the population of interest.

9. **Statisticians are "frightened" of negative numbers**
Most statistical techniques are based upon differences and as such, many of these differences are negative. Unfortunately, if you add together a bunch of positive and negative differences, many of them will cancel each other out. To avoid this, a number of statistical techniques use the trick of squaring the differences first, before the summation, and then square-rooting the result. A classic example occurs in the calculation of the standard deviation. This often leads people to comment that statisticians seem "frightened" of negative numbers.

Once you have understood and accepted the above truths about mathematics (and mathematicians), it should hopefully become clearer to you why it is that we mathematicians do certain things. Equally, these truths should help you to unlock the secret to your success in learning the techniques of mathematics and statistics.

http://www.yourstatsguru.com/the-secret-to-success-in-math/

Maths and Magic

by Rob Eastaway

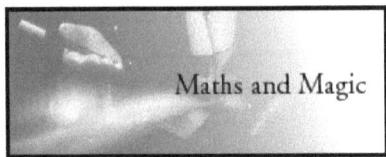

Think of a number (and don't forget it). Double it. Add six. Divide your answer by two. Now take away the number you first thought of. The number in your head is now... three!

Magic? Well, it is to an eight year old. Until you understand the basics of functions and algebra, the thought that a number can be predicted is a surprising one. And of course "magic" and "being surprised" are often the same thing. Pulling a rabbit out of a hat is magic because it goes against what we expect, and also because we can't explain how it has been done.

Let's look at another example of mathematical magic. This trick is going to make a number you choose appear six times (to get the best effect it helps if you have a calculator). Think of a number between 1 and 9. Now multiply it by 7, then by 3, next by 11, then by 37, and finally by 13.

If you haven't seen it before, the result will surprise you and make you smile. And even adults have been know to regard this as a magic trick (especially when it's dressed up with a bit of appropriate patter).

Like all tricks, it has a perfectly logical explanation. The numbers 3, 7, 11, 13 and 37 are the prime factors of 111,111. Why does it appear magical? Because we like pretty patterns, and our experience tells us that multiplying lots of familiar "boring" numbers doesn't normally produce something pretty. Incidentally, numbers that are made up entirely of ones are known by mathematicians as "repunits", and repunits have many interesting properties. For example, $111^2=12321$. Half of all repunits are exactly divisible by 11, and the other half when divided by 11 give a remainder of 1. (Actually that result is pretty obvious when you think about it.) Because of its everyday factors, I find the six digit repunit the most interesting one of all.

Maths and magic have been partners for a long time. Back in the days of Pythagoras, numbers were connected more with mysticism than with conjuring, but discoveries like the "3, 4, 5" triangle were enough to make people believe that some numbers must have magical powers. In the 19th century, Lewis Carroll (a.k.a. Charles Dodgson, a maths lecturer at Oxford) was fascinated by all sorts of tricks and puzzles to do with numbers, some of which magicians still use today. And in modern times, the maths populariser Martin Gardner is one of many mathematicians who are also practising conjurers.

Lewis Carroll

All of these mathemagicians trade off the fact that you can usually predict precisely the outcome of doing something in mathematics, but only if you know the secret beforehand. And since so few people know the secrets of maths, it provides rich possibilities for mind-reading and other "miraculous" deeds.

One of the most ancient of mathematical curiosities is the so-called magic square. This one was known to the ancient Chinese, among others:

$$\begin{array}{ccc} 8 & 1 & 6 \\ 3 & 5 & 7 \\ 4 & 9 & 2 \end{array}$$

This square has the interesting property that every row and diagonal add up to 15, and although you can change it by rotating or reflecting it, the basic arrangement is unique. Unfortunately it is too well known to rate as a good bit of magic, but there are variants which are less well known and therefore more magical.

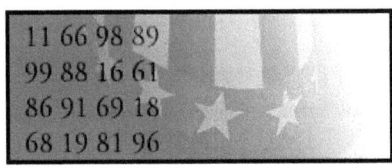

For example, I like this simple square:

$$\begin{array}{cccc} 11 & 66 & 98 & 89 \\ 99 & 88 & 16 & 61 \\ 86 & 91 & 69 & 18 \\ 68 & 19 & 81 & 96 \end{array}$$

In the square, every row, column and diagonal adds to 264. If you study it a bit more closely you should be able to see the simple principle behind it. However, what makes it particularly magical is that if you turn the square upside down, it becomes what appears to be a different magic square, but with the same magic total.

(In fact all of the numbers on the original square are simply mapped onto new positions in the inverted square, though it takes a while to figure out what the rule is for which goes where.)

Wouldn't it be remarkable if a magic square could remain magic not only when upside down but also when viewed in a mirror? Astonishingly, there is one such square. It has digits which when rotated and reflected still stay as legitimate digits. To achieve this, the numbers have to be written in the style of those found on a calculator display. This multi-symmetrical magic square is four-by-four like the one above, with an identical pattern but with some of the digits changed. I'm not going to reveal it, because part of the magic is discovering it for yourself.

I started with an example of a think-of-a-number trick which a child regards as magic but an adult normally regards as a "so what?". Curiously, it can sometimes work the other way around. Adults can be surprised by things that children regard as unexceptional.

Here is an example. Let's suppose that the world is a perfect sphere and you have tied a piece of string tightly around it, so tightly that you can't even squeeze a razor blade underneath. Now cut the string and add in an extra metre to it. Compared to the enormous length of string around the earth, you have only inserted a tiny bit of slack. So the question is, how much slack is there? If millions of people spread out all along the string now tried to lift up the string at the same time, would there be enough slack for them all to squeeze a razor blade underneath? Could they possibly even get their fingers under it?

An adult's intuition usually says that even the razor blades would struggle to get through. Which makes the real answer gobsmacking. It turns out that around the earth there would now be enough slack to let millions of rabbits get under the string without even having to squeeze. The answer is so surprising (if you haven't heard it before) that it seems impossible - or magical, depending on your point of view.

To work out why it is true, you just need some simple algebra.

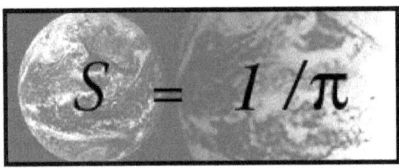

$$S = 1/\pi$$

Let's call the diameter of the earth D, so its circumference C is $\pi \times D$.

We add in a metre of string, to make the new circumference $C_{\text{new}} = \pi \times (D + S)$, where S is the extra diameter (or slack) in the string. But we know that $C_{\text{new}} = C + 1$, so $\pi \times D + 1 = \pi \times D + \pi \times S$.

Cancel out the $\pi \times D$ to give $1 = \pi \times S$, or $S = 1/\pi$, which is roughly 32cm. So the extra diameter is about 32cm, meaning the extra radius is 16cm. In other words, the circle of string will now clear the earth by 16cm all the way round. That's more than enough space for the all rabbits to crawl through.

But young children don't have the same reaction to this "trick", mainly because their sense of the "right" sort of answer to expect is not well enough developed. In fact to them it would be disappointing if the rabbits *couldn't* squeeze under the string!

It reminds me of a story told in the book *Sophie's World*. Mum, Dad and two year old Thomas are at breakfast. Suddenly, Dad flies up and floats around the ceiling. Thomas smiles as he points and says "Look, Daddy's flying". Mum screams and drops the jam.

This simple example demonstrates something that happens throughout the world of maths. Something is only magic if it goes against what your experience tells you to expect. What the two-year-old Thomas saw was exciting, but no more magical than countless other new experiences he saw every day (like the fact that when you drop a bottle it smashes into hundreds of pieces), whereas what his mother saw went against everything she thought she knew.

In the same way, many maths tricks are only surprising to mathematicians who have spent years encountering results that have led them to expect something else. That's why sometimes clever people can be the easiest to fool.

So magical maths isn't just limited to the minds of primary school children. However far you go in the exploration of this subject, you can be certain that there will be things around the corner waiting to surprise you.

About the author

Rob Eastaway is an independent lecturer, and a consultant to the Millennium Maths Project. He specialises in the everyday applications of mathematics. His books include "Why do buses come in threes?", "Guinness Book of Mindbenders" and "The Memory Kit".

http://plus.maths.org/content/maths-and-magic

Pythagorean Mathematics

CONCERNING the secret significance of numbers there has been much speculation. Though many interesting discoveries have been made, it may be safely said that with the death of Pythagoras the great key to this science was lost. For nearly 2500 years philosophers of all nations have attempted to unravel the Pythagorean skein, but apparently none has been successful. Notwithstanding attempts made to obliterate all records of the teachings of Pythagoras, fragments have survived which give clues to some of the simpler parts of his philosophy. The major secrets were never committed to writing, but were communicated orally to a few chosen disciples. These apparently dated not divulge their secrets to the profane, the result being that when death sealed their lips the arcana died with diem.

Certain of the secret schools in the world today are perpetuations of the ancient Mysteries, and although it is quite possible that they may possess some of the original numerical formulæ, there is no evidence of it in the voluminous writings which have issued from these groups during the last five hundred years. These writings, while frequently discussing Pythagoras, show no indication of a more complete knowledge of his intricate doctrines than the post-Pythagorean Greek speculators had, who talked much, wrote little, knew less, and concealed their ignorance under a series of mysterious hints and promises. Here and there among the literary products of early writers are found enigmatic statements which they made no effort: to interpret. The following example is quoted from Plutarch:

"The Pythagoreans indeed go farther than this, and honour even numbers and geometrical diagrams with the names and titles of the gods. Thus they call the equilateral triangle head-born Minerva and Tritogenia, because it may be equally divided by three perpendiculars drawn from each of the angles. So the unit they term Apollo, as to the number two they have affixed the name of strife and audaciousness, and to that of three, justice. For, as doing an injury is an extreme on the one side, and suffering one is an extreme on the on the one side, and suffering in the middle between them. In like manner the number thirty-six, their Tetractys, or sacred Quaternion, being composed of the first four odd numbers added to the first four even ones, as is commonly reported, is looked upon by them as the most solemn oath they can take, and called Kosmos." (*Isis and Osiris.*)

Earlier in the same work, Plutarch also notes: "For as the power of the triangle is expressive of the nature of Pluto, Bacchus, and Mars; and the properties of the square of Rhea, Venus, Ceres, Vesta, and Juno; of the Dodecahedron of Jupiter; so, as we are informed by Eudoxus, is the figure of fifty-six angles expressive of the nature of Typhon." Plutarch did not pretend to explain the inner significance of the symbols, but believed that the relationship which Pythagoras established between the geometrical solids and the gods was the result of images the great sage had seen in the Egyptian temples.

Albert Pike, the great Masonic symbolist, admitted that there were many points concerning which he could secure no reliable information. In his *Symbolism*, for the 32° and 33°, he wrote: "I do not understand why the 7 should be called Minerva, or the cube, Neptune." Further on he added: "Undoubtedly the names given by the Pythagoreans to the different numbers were themselves enigmatical and symbolic-and there is little doubt that in the time of Plutarch the meanings these names concealed were lost. Pythagoras had succeeded too well in concealing his symbols with a veil that was from the first impenetrable, without his oral explanation * * *."

This uncertainty shared by all true students of the subject proves conclusively that it is unwise to make definite statements founded on the indefinite and fragmentary information available concerning the Pythagorean system of mathematical philosophy. The material which follows represents an effort to collect a few salient points from the scattered records preserved by disciples of Pythagoras and others who have since contacted his philosophy.

METHOD OF SECURING THE NUMERICAL POWER OF WORDS

The first step in obtaining the numerical value of a word is to resolve it back into its original tongue. Only words of Greek or Hebrew derivation can be successfully analyzed by this method, and *all words must be spelled in their most ancient and complete forms.* Old Testament words and names, therefore, must be translated back into the early Hebrew characters and New Testament words into the Greek. Two examples will help to clarify this principle.

The *Demiurgus* of the Jews is called in English *Jehovah*, but when seeking the numerical value of the name *Jehovah* it is necessary to resolve the name into its Hebrew letters. It becomes הוהי, and is read from right to left. The Hebrew letters are: ה, He; ו, Vau; ה, He; י, Yod; and when reversed into the English order from left to right read: *Yod-He-Vau-He*. By consulting the foregoing table of letter values, it is found that the four characters of this sacred name have the following numerical significance: *Yod* equals 10. *He* equals 5, *Vau* equals 6, and the second *He* equals 5. Therefore, 10+5+6+5=26, a synonym of *Jehovah*. If the English letters were used, the answer obviously would not be correct.

The second example is the mysterious Gnostic pantheos *Abraxas*. For this name the Greek table is used. Abraxas in Greek is Ἀβραξας. Α = 1, β = 2, ρ = 100, α = 1, ξ =60, α = 1, ς = 200, the sum being 365, the number of days in the year. This number furnishes the key to the mystery of Abraxas, who is symbolic of the 365 Æons, or Spirits of the Days, gathered together in one composite personality. *Abraxas* is symbolic of five creatures, and as the circle of the year actually consists of 360 degrees, each of the emanating deities is one-fifth of this power, or 72, one of the most sacred numbers in the Old Testament of the Jews and in their Qabbalistic system. This same method is used in finding the numerical value of the names of the gods and goddesses of the Greeks and Jews.

All higher numbers can be reduced to one of the original ten numerals, and the 10 itself to 1. Therefore, all groups of numbers resulting from the translation of names of deities into their numerical equivalents have a basis in one of the first ten numbers. By this system, in which the digits are added together, 666 becomes 6+6+6 or 18, and this, in turn, becomes 1+8 or 9. According to Revelation, 144,000 are to be saved. This number becomes 1+4+4+0+0+0, which equals 9, thus proving that both the Beast of Babylon and the number of the saved refer to man himself, whose symbol is the number 9. This system can be used successfully with both Greek and Hebrew letter values.

The original Pythagorean system of numerical philosophy contains nothing to justify the practice now in vogue of changing the given name or surname in the hope of improving the temperament or financial condition by altering the name vibrations.

There is also a system of calculation in vogue for the English language, but its accuracy is a matter of legitimate dispute. It is comparatively modern and has no relationship either to the Hebrew Qabbalistic system or to the Greek procedure. The claim made by some that it is Pythagorean is not supported by any tangible evidence, and there are many reasons why such a contention is untenable. The fact that Pythagoras used 10 as the basis of calculation, while this system uses 9--an imperfect number--is in itself almost conclusive. Furthermore, the arrangement of the Greek and Hebrew letters does not agree closely enough with the English to permit the application of the number sequences of one language to the number sequences of the others. Further experimentation with

SECRETS & TRICKS OF MATHEMATICS

THE NUMERICAL VALUES OF THE HEBREW, GREEK, AND SAMARITAN ALPHABETS.

From Higgins' *Celtic Druids*.

Column
1. Names of the Hebrew letters.
2. Samaritan Letters.
3. Hebrew and Chaldean letters.
4. Numerical equivalents of the letters.
5. Capital and small Greek letters.
6. The letters marked with asterisks are those brought to Greece from Phœnicia by Cadmus.
7. Name of the Greek letters.
8. Nearest English equivalents to the Hebrew, Greek, and Samaritan Letters.

NOTE. When used at the end of a word, the Hebrew *Tau* has the numerical value 440, *Caph* 500, *Mem* 600, *Nun* 700, *Pe* 800, *Tzadi* 900. A dotted *Alpha* and a dashed *Aleph* have the value of 1,000.

the system may prove profitable, but it is without basis in antiquity. The arrangement of the letters and numbers is as follows:

1	2	3	4	5	6	7	8	9
A	B	C	D	E	F	G	H	I
J	K	L	M	N	O	P	Q	R
S	T	U	V	W	X	Y	Z	

The letters under each of the numbers have the value of the figure at: the top of the column. Thus, in the word *man*, $M = 4$, $A = 1$, $N = 5$: a total of 10. The values of the numbers are practically the same as those given by the Pythagorean system.

AN <u>INTRODUCTION</u> TO THE PYTHAGOREAN THEORY OF NUMBERS

(The following outline of Pythagorean mathematics is a paraphrase of the opening chapters of Thomas Taylor's *Theoretic Arithmetic*, the rarest and most important compilation of Pythagorean mathematical fragments extant.)

The Pythagoreans declared arithmetic to be the mother of the mathematical sciences. This is proved by the fact that geometry, music, and astronomy are dependent upon it but it is not dependent upon them. Thus, geometry may be removed but arithmetic will remain; but if arithmetic be removed, geometry is eliminated. In the same manner music depends upon arithmetic, but the elimination of music affects arithmetic only by limiting one of its expressions. The Pythagoreans also demonstrated arithmetic to be prior to astronomy, for the latter is dependent upon both geometry and music. The size, form, and motion of the celestial bodies is determined by the use of geometry; their harmony and rhythm by the use of music. If astronomy be removed, neither geometry nor music is injured; but if geometry and music be eliminated, astronomy is destroyed. The priority of both geometry and music to astronomy is therefore established. Arithmetic, however, is prior to all; it is primary and fundamental.

Pythagoras instructed his disciples that the science of mathematics is divided into two major parts. The first is concerned with the *multitude*, or the constituent parts of a thing, and the second with the *magnitude*, or the relative size or density of a thing.

Magnitude is divided into two parts--magnitude which is stationary and magnitude which is movable, the stationary pare having priority. *Multitude* is also divided into two parts, for it is related both to itself and to other things, the first relationship having priority. Pythagoras assigned the science of arithmetic to multitude related to itself, and the art of music to multitude related to other things. Geometry likewise was assigned to stationary magnitude, and spherics (used partly in the sense of astronomy) to movable magnitude. Both multitude and magnitude were circumscribed by the circumference of mind. The atomic theory has proved size to be the result of number, for a mass is made up of minute units though mistaken by the uninformed for a single simple substance.

Owing to the fragmentary condition of existing Pythagorean records, it is difficult to arrive at exact definitions of terms. Before it is possible, however, to unfold the subject further some light must he cast upon the meanings of the words number, monad, and one.

The *monad* signifies (a) the all-including ONE. The Pythagoreans called the monad the "noble number, Sire of Gods and men." The monad also signifies (b) the sum of any combination of numbers considered as a whole. Thus, the universe is considered as a monad, but the individual parts of the universe (such as the planets and elements) are monads in relation to the parts of which they themselves are composed, though they, in turn, are parts of the greater monad formed of their sum. The monad may also be likened (c) to the seed of a tree which, when it has grown, has many branches (the numbers). In other words, the numbers are to the monad what the branches of the tree are to the seed of the tree. From the study of the mysterious Pythagorean monad, Leibnitz evolved his magnificent theory of the world atoms--a theory in perfect accord with the ancient teachings of the Mysteries, for Leibnitz himself was an initiate of a secret school. By some Pythagoreans the monad is also considered (d) synonymous with the *one*.

Number is the term applied to all numerals and their combinations. (A strict interpretation of the term number by certain of the Pythagoreans excludes 1 and 2.) Pythagoras defines *number* to be the extension and energy of the spermatic reasons contained in the monad. The followers of Hippasus declared number to be the first pattern used by the Demiurgus in the formation of the universe.

The *one* was defined by the Platonists as "the summit of the many." The *one* differs from the monad in that the term *monad* is used to designate the sum of the parts considered as a unit, whereas the *one* is the term applied to each of its integral parts.

There are two orders of number: *odd* and *even*. Because unity, or 1, always remains indivisible, the odd number cannot be divided equally. Thus, 9 is 4+1+4, the unity in the center being indivisible. Furthermore, if any odd number be divided into two parts, one part will always be odd and the other even. Thus, 9 may be 5+4, 3+6, 7+2, or 8+1. The Pythagoreans considered the odd number--of which the monad was the prototype--to be definite and masculine. They were not all agreed, however, as to the nature of unity, or 1. Some declared it to be positive, because if added to an even (negative) number, it produces an odd (positive) number. Others demonstrated that if unity be added to an odd number, the latter becomes even, thereby making the masculine to be feminine. Unity, or 1, therefore, was considered an androgynous number, partaking of both the masculine and the feminine attributes; consequently both odd and even. For this reason the Pythagoreans called it *evenly-odd*. It was customary for the Pythagoreans to offer sacrifices of an uneven number of objects to the superior gods, while to the goddesses and subterranean spirits an even number was offered.

Any even number may be divided into two equal parts, which are always either both odd or both even. Thus, 10 by equal division gives 5+5, both odd numbers. The same principle holds true if the 10 be unequally divided. For example, in 6+4, both parts are even; in 7+3, both parts are odd; in 8+2, both parts are again even; and in 9+1, both parts are again odd. Thus, in the even number, however it may be divided, the parts will always be both odd or both even. The Pythagoreans considered the even number-of which the *duad* was the prototype--to be indefinite and feminine.

The odd numbers are divided by a mathematical contrivance--called "the Sieve of Eratosthenes"--into three general classes: *incomposite, composite,* and *incomposite-composite*.

The *incomposite* numbers are those which have no divisor other than themselves and unity, such as 3, 5, 7, 11, 13, 17, 19, 23, 29, 31, 37, 41, 43, 47, and so forth. For example, 7 is divisible only by 7, which goes into itself once, and unity, which goes into 7 seven times.

The *composite* numbers are those which are divisible not only by themselves and unity but also by some other number, such as 9, 15, 21, 25, 27, 33, 39, 45, 51, 57, and so forth. For example, 21 is divisible not only by itself and by unity, but also by 3 and by 7.

The *incomposite-composite* numbers are those which have no common divisor, although each of itself is capable of division, such as 9 and 25. For example, 9 is divisible by 3 and 25 by 5, but neither is divisible by the divisor of the other; thus they have no common divisor. Because they have individual divisors, they are called composite; and because they have no common divisor, they are called in, composite. Accordingly, the term*incomposite-composite* was created to describe their properties.

Even numbers are divided into three classes: *evenly-even*, *evenly-odd*, and *oddly-odd*.

The *evenly-even* numbers are all in duple ratio from unity; thus: 1, 2, 4, 8, 16, 32, 64, 128, 256, 512, and 1,024. The proof of the perfect *evenly-even* number is that it can be halved and the halves again halved back to unity, as 1/2 of 64 = 32; 1/2 of 32 = 16; 1/2 of 16 = 8; 1/2 of 8 = 4; 1/2 of 4 = 2; 1/2 of 2 = 1; beyond unity it is impossible to go.

The *evenly-even* numbers possess certain unique properties. The sum of any number of terms but the last term is always equal to the last term minus one. For example: the sum of the first and second terms (1+2) equals the third term (4) minus one; or, the sum of the first, second, third, and fourth terms (1+2+4+8) equals the fifth term (16) minus one.

In a series of *evenly-even* numbers, the first multiplied by the last equals the last, the second multiplied by the second from the last equals the last, and so on until in an odd series one number remains, which multiplied by itself equals the last number of the series; or, in an even series two numbers remain, which multiplied by each other give the last number of the series. For example: 1, 2, 4, 8, 16 is an odd series. The first number (1) multiplied by the last number (16) equals the last number (16). The second number (2) multiplied by the second from the last number (8) equals the last number (16). Being an odd series, the 4 is left in the center, and this multiplied by itself also equals the last number (16).

SECRETS & TRICKS OF MATHEMATICS

The *evenly-odd* numbers are those which, when halved, are incapable of further division by halving. They are formed by taking the odd numbers in sequential order and multiplying them by 2. By this process the odd numbers 1, 3, 5, 7, 9, 11 produce the evenly-odd numbers, 2, 6, 10, 14, 18, 22. Thus, every fourth number is evenly-odd. Each of the even-odd numbers may be divided once, as 2, which becomes two 1's and cannot be divided further; or 6, which becomes two 3's and cannot be divided further.

Another peculiarity of the evenly-odd numbers is that if the divisor be odd the quotient is always even, and if the divisor be even the quotient is always odd. For example: if 18 be divided by 2 (an even divisor) the quotient is 9 (an odd number); if 18 be divided by 3 (an odd divisor) the quotient is 6 (an even number).

The evenly-odd numbers are also remarkable in that each term is one-half of the sum of the terms on either side of it. For example: [paragraph continues]

SECRETS & TRICKS OF MATHEMATICS

THE SIEVE OF ERATOSTHENES.

Redrawn from Taylor's *Theoretic Arithmetic*.

This sieve is a mathematical device originated by Eratosthenes about 230 B.C. far the purpose of segregating the composite and incomposite odd numbers. Its use is extremely simple after the theory has once been mastered. All the odd numbers are first arranged in their natural order as shown in the second panel from the bottom, designated *Odd Numbers*. It will then be seen that every third number (beginning with 3) is divisible by 3, every fifth number (beginning with 5;) is divisible by 5, every seventh number (beginning with 7) is divisible by 7, every ninth number (beginning with 9) is divisible by 9, every eleventh number (beginning with 11) is divisible by 11, and so on to infinity. This system finally sifts out what the Pythagoreans called the "incomposite" numbers, or those having no divisor other than themselves and unity. These will be found in the lowest panel, designated *Primary and Incomposite Numbers*. In his *History of Mathematics*, David Eugene Smith states that Eratosthenes was one of the greatest scholars of Alexandria and was called by his admirers "the second Plato." Eratosthenes was educated at Athens, and is renowned not only for his sieve but for having computed, by a very ingenious method, the circumference and diameter of the earth. His estimate of the earth's diameter was only 50 miles less than the polar diameter accepted by modern scientists. This and other mathematical achievements of Eratosthenes, are indisputable evidence that in the third century before Christ the Greeks not only knew the earth to be spherical in farm but could also approximate, with amazing accuracy, its actual size and distance from both the sun and the moon. Aristarchus of Samos, another great Greek astronomer and mathematician, who lived about 250 B.C., established by philosophical deduction and a few simple scientific instruments that the earth revolved around the sun. While Copernicus actually believed himself to be the discoverer of this fact, he but restated the findings advanced by Aristarchus seventeen hundred years earlier.

[paragraph continues] 10 is one-half of the sum of 6 and 14; 18 is one-half the sum of 14 and 22; and 6 is one-half the sum of 2 and 10.

The oddly-odd, or unevenly-even, numbers are a compromise between the evenly-even and the evenly-odd numbers. Unlike the evenly-even, they cannot be halved back to unity; and unlike the evenly-odd, they are capable of more than one division by halving. The oddly-odd numbers are formed by multiplying the evenly-even numbers above 2 by the odd numbers above one. The odd numbers above one are 3, 5, 7, 9, 11, and so forth. The evenly-even numbers above 2 are 4, 8, 16, 32, 64, and soon. The first odd number of the series (3) multiplied by 4 (the first evenly-even number of the series) gives 12, the first oddly-odd number. By multiplying 5, 7, 9, 11, and so forth, by 4, oddly-odd numbers are found. The other oddly-odd numbers are produced by multiplying 3, 5, 7, 9, 11, and so forth, in turn, by the other evenly-even numbers (8, 16, 32, 64, and so forth). An example of the halving of the oddly-odd number is as follows: 1/2 of 12 = 6; 1/2 of 6 = 3, which cannot be halved further because the Pythagoreans did not divide unity.

SECRETS & TRICKS OF MATHEMATICS

Even numbers are also divided into three other classes: *superperfect, deficient,* and *perfect.*

Superperfect or *superabundant* numbers are such as have the sum of their fractional parts greater than themselves. For example: 1/2 of 24 = 12; 1/4 = 6; 1/3 = 8; 1/6 = 4; 1/12 = 2; and 1/24 = 1. The sum of these parts (12+6+8+4+2+1) is 33, which is in excess of 24, the original number.

Deficient numbers are such as have the sum of their fractional parts less than themselves. For example: 1/2 of 14 = 7; 1/7 = 2; and 1/14 = 1. The sum of these parts (7+2+1) is 10, which is less than 14, the original number.

Perfect numbers are such as have the sum of their fractional parts equal to themselves. For example: 1/2 of 28 = 14; 1/4 = 7; 1/7 = 4; 1/14 = 2; and 1/28 = 1. The sum of these parts (14+7+4+2+1) is equal to 28.

The perfect numbers are extremely rare. There is only one between 1 and 10, namely, 6; one between 10 and 100, namely, 28; one between 100 and 1,000, namely, 496; and one between 1,000 and 10,000, namely, 8,128. The perfect numbers are found by the following rule: The first number of the evenly-even series of numbers (1, 2, 4, 8, 16, 32, and so forth) is added to the second number of the series, and if an incomposite number results it is multiplied by the last number of the series of evenly-even numbers whose sum produced it. The product is the first perfect number. For example: the first and second evenly-even numbers are 1 and 2. Their sum is 3, an incomposite number. If 3 be multiplied by 2, the last number of the series of evenly-even numbers used to produce it, the product is 6, the first perfect number. If the addition of the evenly-even numbers does not result in an incomposite number, the next evenly-even number of the series must be added until an incomposite number results. The second perfect number is found in the following manner: The sum of the evenly-even numbers 1, 2, and 4 is 7, an incomposite number. If 7 be multiplied by 4 (the last of the series of evenly-even numbers used to produce it) the product is 28, the second perfect number. This method of calculation may be continued to infinity.

Perfect numbers when multiplied by 2 produce superabundant numbers, and when divided by 2 produce deficient numbers.

The Pythagoreans evolved their philosophy from the science of numbers. The following quotation from Theoretic Arithmetic is an excellent example of this practice:

"Perfect numbers, therefore, are beautiful images of the virtues which are certain media between excess and defect, and are not summits, as by some of the ancients they were supposed to be. And evil indeed is opposed to evil, but both are opposed to one good. Good, however, is never opposed to good, but to two evils at one and the same time. Thus timidity is opposed to audacity, to both [of] which the want of true courage is common; but both timidity and audacity are opposed to fortitude. Craft also is opposed to fatuity, to both [of] which the want of intellect is common; and both these are opposed to prudence. Thus, too, profusion is opposed to avarice, to both [of] which illiberality is common; and both these are opposed to liberality. And in a similar manner in the other virtues; by all [of] which it is evident that perfect numbers have a great similitude to the virtues. But they also resemble the virtues on another account; for they are rarely found, as being few, and they are generated in a very constant order. On the contrary, an infinite multitude of superabundant and diminished numbers may be found, nor are they disposed in any orderly series, nor generated from any certain end; and hence they have a great similitude to the vices, which are numerous, inordinate, and indefinite."

http://www.sacred-texts.com/eso/sta/sta16.htm

11 Super Badass Math Tricks

written by Sam Greenspan

Look. We all know math tricks aren't super badass. You don't sit around with your friends comparing math tricks. (And if you do sit around with your friends comparing math tricks, you guys are probably far more advanced than the math tricks in this list. RDRR forever, fellas.)

But this website is a private experience. It's just you, me and whatever software your company put on your computer to spy on you. So if you want to test these out... punch a few numbers into a calculator... jot a few things down on a piece of paper... or simply look up and off to the side as you deeply concentrate on mental math -- it'll be our secret. It's just between you, me and the keylogger. And possibly the window washer behind you.And the guy who's walking around from desk to desk right now saying "Mondays, huh?"

That's what we call the Dennis Miller ratio. Glaven!

1 The Trick Of The 11s. Since this is 11 Points and all, I felt like I should orient the first trick around the number 11. (I know some sensitive types might object to using the word "orient" in a math list, and for that I apologize.)

This is a simple multiplication trick that will allow you to multiply by 11 at will. And it's not "multiply the number by 10 and then add it to that result." Even though that works too. But that's a cop out. That's like losing weight through diet and exercise and not by doing a crazy diet like the one where you take pregnancy hormones and eat 500 calories a day.

When multiplying a two digit number by 11, the result is always first digit of the number, sum of the two numbers, last digit. Example: 11 x 18 is 198. 11 x 32 is 352. If the sum of the digits comes out greater than nine, add one to the first digit. 11 x 78 is 858.

SECRETS & TRICKS OF MATHEMATICS

So if someone said "What website is 64 times better than 11 Points" you'd know it's the website 704 Points without even busting out a calculator. (Also, if anyone does want to start a website featuring 704-item lists, I'd strongly recommend against it. Trying to get to 11 is enough to make a grown man cry.)

2 At What Age Should You Get Married? I like this one. It uses Euler's number (e, which roughly equals 2.71828) to tell you the age where you have the highest probability of getting married to the ideal person.

1. Figure out the age range when you consider yourself marriage eligible. If you're already well into it, let's say 22 to 45. (If you're in Wisconsin or Minnesota, go ahead and switch that down to 18 to 24. If you're in a major city, feel free to pump it up to 31 to 86.)
2. Subtract the oldest age from the youngest age to calculate your total number of marryin' years.
3. Divide that number by e, or 2.71828.
4. Add the result to your youngest marrying age.

So if you've set a range of 22 to 45, that's 23 years. Divided by Eular's number is 8.46.

Added to 22 is 30.46, meaning that between 30 and 31 is the time when you're at your statistical peak (and *ready to stop messing around with data points and settle down with that one special data point*).

SECRETS & TRICKS OF MATHEMATICS

3. 111 Birthdays Of Joy. This is a trick I may or may not have been turned on to while watching the "TMZ" TV show. (I have a girlfriend. To those single guys out there who wonder what it's like to have a girlfriend, the 11-word summary is "watching terrible TV, some of which you begrudgingly grow to like.")

Now, on to the trick...

1. Take the last two digits of the year you were born.
2. Add that to the age you're going to turn (or you've already turned) here in 2011.

The result should be... 111. Unless you were born before 1900 or after 1999. In either case, you shouldn't be reading this website. You should be telling your life story to a biographer or putting down the piece of cake to go play outside, respectively.

It's called "Gold Case".

4 Deal... Or No Deal. I'm not sure if "Deal Or No Deal" will ever be back on TV. After all, it made "Wheel Of Fortune" look like a Mensa exam. I did not begrudgingly grow to enjoy watching it. But if it does come back -- or if you still play the video game versions or arcade version to win prize tickets or something -- here's the quick and dirty math.

You should never take the banker's offer unless it's greater than the simple mean of the remaining briefcases. So if the cases you have left are $5, $50, $250, $10,000, and $400,000, you should only take an offer that's more than $82,061. Yes, even though only one of the five cases in play has that amount.

Forget risk, forget going home empty handed, forget it. ***Take your brain and risk tolerance out of it and put your faith in math.*** That way you have something to blame. And if it screws you, just breathe a bunch of germs onto Howie Mandel.

5 The 1001 Cloning Machine. I tried to give all the fairly sterile math tricks on this list ornate and alluring names. I think I had to work the hardest for this one. Because this is... something.

1. Take any three digit number.
2. Multiply it by 1,001.

The result is always the same: A six-digit number that's your original number, twice in a row. 456 x 1001 is 456,456. And once you get a few answers and prove to yourself that this is true, ***jump up, point at the sky, and yell "Scheherazade!"*** Don't worry. Everyone will know what you're talking about.

6 The Nine Folding Procedure. If you read this site, and specifically a math list on this site, you know the product of nine times any single digit number without having to think about it. So teach this trick to someone who doesn't.

To figure out the result, **hold up your hands, then fold down the finger that corresponds to the number you're multiplying by nine.** So if it's 9 x 6, hold down your sixth finger (aka the thumb on your right hand). Then take a look at what you've got. Five fingers before the fold, four fingers after... the product is 54.

I remember learning this in early elementary school, then forgetting it, then remembering it again yesterday when I was researching this list. That's quite the 25-year layoff. There's a montage playing in my head of all the things that happened in the time between when I learned the nine folding procedure and when I remembered it. "Mr. Gorbachev, tear down this wall!" "I'm so excited, I'm so scared!" "I did not have sexual relations with that woman." "A white Ford Bronco" and so on. All the way up to "Hide ya kids, hide ya wife."

Photoshop replace color.

7 **Imaginary Taxonimy.** This is a multi-step trick, so let's all do it together...

1. Think of a number one through 10.
2. Multiply it by nine. Feel free to fold down your fingers if the pressure's getting to you.
3. Add the digits together.
4. Subtract five from that sum.
5. Find the letter of the alphabet that corresponds to that number (A=1, B=2, C=3, D=4, etc.)
6. Think of a country that starts with that letter.
7. Think of an animal that starts with the last letter of your country.
8. Think of a color that starts with the last letter of your animal.

According to The Internet, this is where I'm supposed to tell you "Come on, man, there aren't any orange kangaroos in Denmark!"

The only problem is that esoteric smart asses like me and you are probably thinking of **amber iguanas from Djibouti** or white cows from the Dominican Republic.

So I think I have to structure this joke better. Instead of subtracting five from the sum there in step four, let's add eight. Now we should all agree on teal, tan or taupe rabbits from Qatar.

8 **Kaprekar's Constantly Saying The Same Damn Thing.** D.R. Kaprekar was an Indian mathematician. (And, coincidentally,

with a January 17th birthday, Kaprekar was a Capricorn.) And while other mathematicians focused on numbers like pi or e, Kaprekar set his sights on... 6,174.

1. Take any four digit number. (Only rule is that it has to contain at least two different numbers, so 5555 would be out, but 5554 would be OK.)
2. Put the digits in descending and ascending order. (So if the number you picked was 7048, you'd have 8740 and 0478.)
3. Subtract the smaller number from the larger number.
4. If your result isn't 6174, repeat steps two and three with your new number.

Within seven tries, you'll end up with Kaprekar's constant. That's right. It's magic. Mathemagician magic, not *wingardiumleviosa* magic, but still magic.

9 The Percentage Fun House Mirror. This one is useful... I don't know, somewhere. ***It's just a good trick if you every find yourself in You Got Served-style battle where instead of dancing you're battling via calculating percentages.***

And it is: *x% of y = y% of x*. In other words, 20 percent of 40 is the same as 40 percent of 20. Let's see you multiply, sucka, you got nothing on me.

SECRETS & TRICKS OF MATHEMATICS

Don't pigeon hole me.

10 No Way, We Have The Same Birthday, That's So Crazy!
If there are 367 people in a room, there's a guarantee that at least two of them will have the same birthday. (And you even account for freakish Leap Day babies.) That's called the Pigeon Hole Principle, which says if you shove n pigeons into $n-1$ holes, at least one hole is guaranteed to have multiple pigeons. I have no idea why the math community is going around shoving pigeons into holes, but I have suspicions it's connected to their inability to hold their liquor.

Also, that definitely feels like one of the analogies Charlie would've used on the show "Numbers" (err... "Numb3rs") to explain to the FBI something about murderers or terrorists or something.

Now here's the birthday twist. It's really not that hard to get two people with the same birthday into the same room. *If there are only 23 people in the room, there's greater than a 50-50 chance that two of them will have the same birthday.* And if there are 41 people in the room, it goes to a 90 percent chance.

SECRETS & TRICKS OF MATHEMATICS

11 Anti-Lychrel Palindrome Syndrome (aka A Racecar Named Desire). Through addition and inversion, you can eventually get almost any number to whittle down to a palindrome. (That is, the same forwards and backwards.) Most of them get there in seven steps or less.

Take a number down, flip it, and reverse it. (Missy Elliott style, you see.) Then add your original number and its mirror image. Repeat until you have a palindrome. Check this out...

1492 + 2941 = 4433
4433 + 3344 = 7777

525600 + 6525 = 532125
532125 + 521235 = 1053360
1053360 + 633501 = 1686861

But there are some numbers that just won't go palindrome. Those are called Lychrel numbers. The smallest is 196. (Others include 295, 394, 493, 592, 689, 691, 788, 790, 879, 887, 978, 986, 1947 and 1997.) So don't waste your time with them. Or you'll be adding and inverting all day. Which is too Sisyphean, even as a time waster in class or at work.

http://www.11points.com/Misc/11_Super_Badass_Math_Tricks

How to Learn Math Formulas

In a recent <u>IntMath Poll</u>, readers indicated that the hardest thing about math was learning the formulas. Here are 10 things you can do to improve your memory for math formulas.

1. Read ahead

Read over tomorrow's math lesson today. Get a general idea about the new formulas in advance, before your teacher covers them in class.

As you read ahead, you will recognize some of it, and other parts will be brand new. That's OK – when your teacher is explaining them you already have a "hook" to hang this new knowledge on and it will make more sense — and it will be easier to memorize the formulas later.

This technique also gives you an overview of the diagrams, graphs and vocabulary in the new section. Look up any new words in a dictionary so you reduce this stumbling block in class.

This step may only take 15 minutes or so before each class, but will make a huge difference to your understanding of the math you are studying.

I always used to read ahead when I was a student and I would be calm in class while all my friends were stressed out and confused about the new topic.

2. Meaning

All of us find it very difficult to learn meaningless lists of words, letters or numbers. Our brain cannot see the connections between the words and so they are quickly forgotten.

Don't just try to learn formulas by themselves — it's just like learning that meaningless list.

When you need to learn formulas, also learn the **conditions** for each formula (it might be something like "if $x > 0$").

Also draw a relevant diagram or graph each time you write the formula (it might be a parabola, or perhaps a circle). You will begin to associate the picture with the formula and then later when you need to recall that formula, the associated image will help you to remember it (and its meaning, and its conditions).

During exams, many of my students would try to answer a question with the wrong formula! I could see that they successfully learned the formula, but they had no idea how to apply it. Diagrams, graphs and pictures always help.

Most of us find it difficult to learn things in a vacuum, so make sure you learn the formulas in their right context.

When you create your summary list of formulas, include conditions and relevant pictures, graphs and diagrams.

3. Practice

You know, math teachers don't give you homework because they are nasty creatures. They do it because they know repetition is a very important aspect of learning. If you practice a new skill, the connections between neurons in your brain are strengthened. But if you don't practice, then the weak bonds are broken.

If you try to learn formulas without doing the practice first, then you are just making it more difficult for yourself.

4. Keep a list of symbols

Most math formulas involve some Greek letters, or perhaps some strange symbols like ^ or perhaps a letter with a bar over the top.

When we learn a foreign language, it's good to keep a list of the new vocabulary as we come across it. As it gets more complicated, we can go back to the list to remind us of the words we learned recently but are hazy about. Learning mathematics symbols should be like this, too.

Keep a list of symbols and paste them up somewhere in your room, so that you can update it easily and can refer to it when needed. Write out the symbol in words, for example: Σ is "sum"; $\int$ is the "integration" symbol and Φ is "capital phi", the Greek letter.

Just like when learning whole formulas, include a small diagram or graph to remind you of where each symbol came from.

Another way of keeping your list is via flash cards. Make use of dead time on the bus and learn a few formulas each day.

5. Absorb the formulas via different channels

I've already talked about writing and visual aids for learning formulas. Also process and learn each one by hearing it and speaking it.

An example here is the formula for the derivative of a fraction involving x terms on the top and bottom (known as the "Quotient Rule"). Then in words, the derivative is:

dy/dx = bottom **times** derivative of top **minus** top **times** derivative of bottom **all over** bottom squared.

The formula is actually as follows, if we let u = numerator and v = denominator of the fraction, then:

$$\frac{dy}{dx} = \frac{v\frac{du}{dx} - u\frac{dv}{dx}}{v^2}$$

6. Use memory techniques

Most people are capable of learning lists of unrelated numbers or words, as long as they use the right techniques. Such techniques can be applied to the learning of formulas as well.

One of these techniques is to create a story around the thing you need to learn. The crazier the story, the better it is because it is easier to remember. If the story is set in some striking physical location, it also helps to remember it later.

7. Know why

In many examinations, they give you a math formula sheet so why do you still need to learn formulas? As mentioned earlier, if students don't know what they are doing, they will choose a formula randomly, plug in the values and hope for the best. This usually has bad outcomes and zero marks.

I encourage you to learn the formulas, even if they are given to you in the exam. The process of learning the conditions for how to use the formula and the associated graphs or diagrams, means that you are more likely to use the correct formula and use it correctly when answering the question. This is also good for future learning, because you have a much better grasp of the basics.

8. Sleep on it

Don't under-estimate the importance of sleep when it comes to remembering things. Deep sleep is a phase during the night where we process what we thought about during the day and this is when more permanent memories are laid down. During REM (rapid eye movement) sleep, we rehearse the new skills and consolidate them.

Avoid cramming your math formulas the night before an exam until late. Have a plan for what you are going to learn and spread it out so that it is not overwhelming.

9. Healthy body, efficient brain

The healthier you are, the less you need to worry about sickness distracting from your learning. Spend time exercising and getting the oxygen flowing in your brain. This is essential for learning.

10. Remove distractions

This one is a problem for those of us that love being on the Internet, or listening to music, or talking to our friends. There are just so many things that distract us from learning what we need to learn.

Turn off all those distractions for a set time each day. You won't die without them. Concentrate on the formulas you need to learn and use all the above techniques.

When you are done, reward yourself with some media time — but only after you have really accomplished something.

http://www.intmath.com/blog/how-to-learn-math-formulas

11 Math Tricks That Will Make Your Life So Much Easier

WALTER HICKEY MAR. 11, 2013, 9:19 AM

$$S = \frac{\pi A k c^3}{2hG}$$

Wikimedia Commons

Imagine you're a trader.

You're on the trading floor, trying to price out if a 15-year bond issued by General Electric will generate the returns needed to placate your investors.

Bad news, your calculator is dead and the trader from Cantor Fitzgerald is readying to signal his buy. What to do?

SECRETS & TRICKS OF MATHEMATICS

Well, if all you need to do is double the investment in five-years, you're in luck.

That's probably not the case, and maybe <u>GE</u> isn't issuing 15-year debt. But we compiled a list of eleven math tricks that might just come in handy at various times in life.

If you have a math trick you'd like us to add, leave it in the comments.

Eric Platt also contributed to this report.

The Rule of 72

Need an easy way to determine how long it will take to double your returns? Simply divide the number 72 by your projected growth rate.

So, if your returns are increasing by 10 percent per year, it will take 7.2 years for them to double in size.

SECRETS & TRICKS OF MATHEMATICS

The Rule of 115

Baby Pictures

If you're more inclined to triple your returns, because you're not as risk averse (or perhaps your time horizon is just a tad bit farther out), simply take the number 115 and divide it by your growth rate. This will give you the amount of time it will take to triple your returns.

So, if your returns are increasing by 10 percent per year, it will take 11.5 years for them to triple in size.

SECRETS & TRICKS OF MATHEMATICS

The Rule of 70

Magyar NemzetiMúzeumTörténetiFényképtára, Budapest

The rule of 70 dictates how long it will take for inflation to halve the value of a dollar. Simply divide 70 by your expected rate of inflation.

For example, if you expect 3 percent inflation, then divide 70 by 3. At that rate, it will take 23.3 years before the value of your money is worth half what it is today.

SECRETS & TRICKS OF MATHEMATICS

Squaring numbers in your head

"You can fondle the cube, but it won't respond."

Squaring large numbers can be a real pain sometimes. But if you're plugging something into a formula, easy mental squaring could be a huge asset.

So say you've got a number, x, that you want to square.

Find "d" the difference between the nearest multiple of ten and x.

Then, multiply (x-d) and (x+d). This should be much easier, because one of the numbers is a multiple of ten. Just add d^2, and you've got your square.

Here's an example. I want to find the square of 84. The nearest multiple of ten is 80, so d is 4.

x+d is 88, x-d is 80.

88 X 80 = 6400 + 640 = 7040. Add 4^2 = 16, and you get 7056.

That process, once you get the hang of it, is much easier than just attacking 84^2 head on.

Converting your salary to an hourly figure

You're a salaried employee and trying to figure out how much that wage earns you an hour, maybe for that part-time job you're considering taking on. Take your salary, drop the last three zeros and then divide by the number two.

So if you earn $40,000, you're left with $20 an hour. Numbers work best if you're only working a 40 hour week.

SECRETS & TRICKS OF MATHEMATICS

Take a repeating fraction and turn it into a decimal

metropilot / flickr

This one is a great party trick, provided you find yourself at a terrible party where other guests are discussing repeating decimals.

Repeating decimals are a pain, and oftentimes basic calculators have trouble getting down to brass tacks by rephrasing them as a fraction.

All you need to do to turn a repeating decimal (0.636363...) into a lovely fraction is:

- Find the number that repeats (63)
- Figure out how many places that number has (2)
- Divide the repeater by a number with the same number of places made up of nines (in this case, 99)

So we know 0.636363... = 63/99 = 7/11.

You can do this with much larger repeating numbers as well. 0.726726... is equal to 726/999 which reduces to 242/333.

Easy division with 7

Flickr

Dividing by 7 is probably the most annoying possible aspect of simple arithmetic. There are relatively simple strategies and mental tricks for all of the other divisors between 1 and 10, by 7 stands alone.

Here's where division comes in. Lets say you wanted to divide 9573 by 7. Let's work from the left.

Start with thousands. So 9/7 = 1 with a remainder of 2. So our first digit is 1.

Since we had a remainder of 2, and the hundreds digit is 5, we next get 25/7 = 3 with a remainder of 4. So our next digit is 3.

We have a remainder of 4, and the tens place we have a 7, so we have 47/7 = 6 with a remainder of 5.

SECRETS & TRICKS OF MATHEMATICS

We have a remainder of 5, and in the ones place we have a 3. So 53/7 = 7, with a remainder of 4.

We remember, then that 4/7 is equal to .571428 repeating.

So 9573 divided by 7 is 1367.571428 repeating.

Ways to remember time

Wikimedia Commons

You can use factorials to remember easy stats about time.

- There are 4! (or 4*3*2*1) hours in a day.
- There are 8! minutes in 4 weeks
- There are 10! seconds in 6 weeks.

SECRETS & TRICKS OF MATHEMATICS

Multiplying by 11

unrealitymag.com

You never know when you'll be pricing out an 11-year fixed income product, so this might come in handy. When multiplying a figure by the number 11, follow this pattern: leave the last and first digits alone, then sum each and every pair of digits next to each other (this makes most sense when seen in example):

1. 4,281 x 11 becomes the following digits: (4)(4+2),(2+8)(8+1)(1) or 47,091

When the sum of a pair is greater than 10, carry that digit to the next left pair (as seen above, where 2+8 was 10)

2. Let's try something harder. 9,621,576,521 x 11 becomes: (9)(9+6),(6+2)(2+1)(1+5),(5+7)(7+6)(6+5),(5+2)(2+1)(1) or 105,837,341,731

SECRETS & TRICKS OF MATHEMATICS

Converting fractions with 7 in the denominator to decimals

wikipaintings

Converting fractions to decimals are usually pretty easy when the number in the denominator is less than ten. The glaring exception is with 7 in the denominator.

The one thing you need to remember in order to divide by 7 is that 1/7 = .142857 repeating. That's the key. If you can remember that, with a little practice, dividing by 7 becomes a breeze.

The next thing that you have to realize is that multiples of that just cycle through the six numbers. For example, 2/7 = .285714 repeating. Notice what just happened?

The numbers always cycle in the same sequence. For 1/7, the cycle starts at 1, the lowest digit. For 2/7, the cycle starts with 2, the second lowest digit. For 3/7, (.428571 repeating) the cycle starts with the 4, the third lowest digit. That's the process.

Asset Allocation by Age

ttp://www.flickr.com/photos/rolymo/

This one really isn't a math trick, so much as it is a rule of thumb...

Don't have a financial planner to walk you through asset allocation? A simple way to find out is to subtract your age from the number 120, the number remaining is the percentage of your portfolio that should be in stocks.

For instance, if you're 50, you should be keeping 70% of your holdings in stocks with the remaining 30 percent in fixed income products.

http://www.businessinsider.com/11-awesome-math-tricks-that-will-make-your-life-so-much-easier-2013-2?op=1

Techniques for Mental Arithmetic

Addition

Grouping

When we add up a list of numbers, it is OK to change the order of the numbers and to perform the addition operations in any order. For instance 1+2+3+4 = 1+3+4+2 = (1+3) + (4+2) = 1 + (3+4+2) = 10. Sometimes we can use this to group numbers to make our additions easier.

3 + 4 + 5 + 6 + 7 = (3 + 7) + (4 + 6) + 5 = 10 + 10 + 5 = 25

Here we have found two pairs which add up to 10, 3 + 7 and 4 + 6. (Work out all the pairs which do this!) By doing these additions first we make the whole calculation easier.

Question 1: Try regrouping the following additions using 10-pairs (pairs of numberswhich add to 10) and then adding them.

a) 5 + 6 + 5 + 4 =
b) 3 + 8 + 1 + 7 + 9 + 2 =
c) 6 + 6 + 3 + 4 + 7 =

SECRETS & TRICKS OF MATHEMATICS

d) 1 + 2 + 3 + 4 + 5 + 6 + 7 + 8 + 9 =
e) 8 + 7 + 8 + 3 + 2 + 6 + 4 =

You can also group larger numbers to make your mental calculations easier. For instance 23 + 45 + 17 = (23 + 17) + 45 = 40 + 45 = 85

Question 2: Try the following mental calculations:

a) 16 + 14 + 13 =
b) 12 + 19 + 8 =
c) 31 + 46 + 19 =
d) 37 + 6 + 13 + 14 =
e) 21 + 22 + 28 + 19 =

Multiplication using Areas

Using this technique, you should be able to learn to multiply any pair of two-digit numbers together in your head. (If you know your tables up to 9 x 9)

First we need to explore the connection between multiplication and calculating areas. The area of a rectangle can be found by multiplying its length by its width:

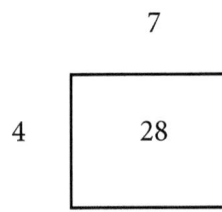

SECRETS & TRICKS OF MATHEMATICS

This area has length 7, width 4 and area 7 x 4 = 28

The good thing is that if a rectangle is broken up into smaller rectangles, it doesn't matter whether the areas of the smaller rectangles are added up to form the total area or whether we add the sides and then multiply to get total area. Look at this example:

```
        4    2
     ┌─────┬───┐
  3  │ 12  │ 6 │
     ├─────┼───┤
     │     │   │
  2  │  8  │ 4 │
     └─────┴───┘
```

The area of this shape can be found by calculating the separate areas of the small rectangles and then adding them:
12 + 6 + 8 + 4 = 30
Or by first finding the total length:

4 + 2 = 6   and the total width:
3 + 2 = 5   and then multiplying to
find the total area: 6 x 5 = 30

Use both of these methods to find the areas of the following rectangles (not drawn to scale). Show your working.

1.

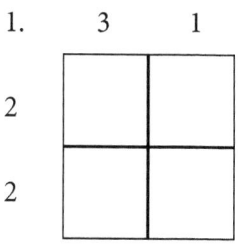

2.   5   3

   4

   3

3.   6   4

   7

   3

4.   6   3

   4

   2

Multiplication of numbers ending with zeros

If you know your tables, you should be able to do simple calculations involving numbers which end with a zero. For instance, if you know that 3 x 7 = 21 then you should be able to see why 30 x 7 = 210 and 3 x 70 = 210. If both of the numbers you are multiplying together end in a 0, then there will be two zeros on the end of the answer. For instance 1 x 1 = 1 and 10 x 10 = 100. Similarly 30 x 70 = 2100

See if you can write down the answers to the following calculations using your tables and the above rules for dealing with end zeros.

1. 2 x 20
2. 4 x 50
3. 70 x 6
4. 20 x 9
5. 8 x 70
6. 9 x 90
7. 30 x 30
8. 50 x 50
9. 40 x 70
10. 60 x 30
11. 80 x 80
12. 90 x 4

Multiplying a two-digit number by a one-digit number

Using a combination of our areas method and the rules above for multiplying numbers ending in a zero, we can simplify a multiplication of a two-digit number by a one-digit number as follows:

7 x 23

Break 23 up as 20 + 3

	20	3
7	140	21

7 x 20 = 140

3 x 7 = 21

Adding together results we get 140 + 20 + 1 = 161, so 7 x 23 = 161

Try doing the following multiplications using this method:

1. 3 x 45
2. 6 x 64
3. 8 x 17
4. 2 x 39
5. 7 x 92
6. 4 x 88
7. 9 x 46
8. 5 x 78

Multiplying a two-digit number by another two-digit number

We use a combination of techniques we have already learnt as follows:

23 x 42 = (20 + 3) x (40 + 2)

	20	3
40	800	120
2	40	6

800 + 120 + 40 + 6 = 966

Try the following:

1. 12 x 17
2. 44 x 52

https://eis.uow.edu.au/content/groups/public/@web/@inf/@math/documents/web/uow062599.pdf

Fast Arithmetic Tips

Mental Calculations - Getting the result fast

1. Addition of 5

When adding 5 to a digit greater than 5, it is easier to first subtract 5 and then add 10.

For example,

7 + 5 = 12.

Also 7 - 5 = 2; 2 + 10 = 12.

2. Subtraction of 5

When subtracting 5 from a number ending with aa digit smaller than 5, it is easier to first add 5 and then subtract 10.

For example,

23 - 5 = 18.

Also 23 + 5 = 28; 28 - 10 = 18.

3. Division by 5

Similarly, it's often more convenient instead to multiply first by 2 and then divide by 10.

For example,

1375/5 = 2750/10 = 275.

More examples and explanation

4. Multiplication by 5

It's often more convenient instead of multiplying by 5 to multiply first by 10 and then divide by 2.

For example,

137×5 = 1370/2 = 685.

More examples and explanation

5. Division by 5

Similarly, it's often more convenient instead to multiply first by 2 and then divide by 10.

For example,

1375/5 = 2750/10 = 275.

More examples and explanation

6. Division/multiplication by 4

Replace either with a repeated operation by 2.

For example,

$124/4 = 62/2 = 31$. Also,

$124 \times 4 = 248 \times 2 = 496$.

7. Division/multiplication by 25

Use operations with 4 instead.

For example,

$37 \times 25 = 3700/4 = 1850/2 = 925$.

More examples and explanation

8. Division/multiplication by 8

Replace either with a repeated operation by 2.

For example,

$124 \times 8 = 248 \times 4 = 496 \times 2 = 992$.

SECRETS & TRICKS OF MATHEMATICS

9. Division/multiplication by 125

Use operations with 8 instead.

For example,

$37 \times 125 = 37000/8 = 18500/4 = 9250/2 = 4625$.

10 Squaring two digit numbers.

i. You should memorize the first 25 squares:

1	2	3	4	5	6	7	8	9	10	11	12	13	14
1	4	9	16	25	36	49	64	81	100	121	144	169	196

15	16	17	18	19	20	21	22	23	24	25
225	256	289	324	361	400	441	484	529	576	625

If you forgot an entry.

ii. Say, you want a square of 13. Do this: add 3 (the last digit) to 13 (the number to be squared) to get $16 = 13 + 3$. Square the last digit: $3^2 = 9$. *Append* the result to the sum: 169.

As another example, find 14^2. First, as before, add the last digit (4) to the number itself (14) to get $18 = 14 + 4$. Next, again as before, square the last digit: $4^2 = 16$. You'd like to append the result (16) to the sum (18) getting 1816 which is clearly too large, for, say, 14 < 20 so that $14^2 < 20^2 = 400$. What you have to do is append 6 and carry 1 to the previous digit (8) making $14^2 = 196$.

More examples and explanation

iii Squares of numbers from 26 through 50.
Let A be such a number. Subtract 25 from A to get x. Subtract x from 25 to get, say, a. Then $A^2 = a^2 + 100x$. For example, if $A = 26$, then $x = 1$ and $a = 24$. Hence $26^2 = 24^2 + 100 = 676$.

More examples and explanation

vi. Squares of numbers from 51 through 99.
If A is between 50 and 100, then $A = 50 + x$. Compute $a = 50 - x$. Then $A^2 = a^2 + 200x$. For example,
$63^2 = 37^2 + 200 \times 13 = 1369 + 2600 = 3969$.

More examples and explanation

11. Any Square.

Assume you want to find 87^2. Find a simple number nearby - a number whose square could be found relatively easy. In the case of 87 we take 90. To obtain 90, we need to add 3 to 87; so now let's subtract 3 from 87. We are getting 84. Finally,

$87^2 = 90 \times 84 + 3^2 = 7200 + 360 + 9 = 7569$.

More examples and explanation

12. Squares Can Be Computed Squentially

In case A is a successor of a number with a known square, you find A⊃ by adding to the latter itself and then A. For example, A = 111 is a successor of a = 110 whose square is 12100. Added to this 110 and then 111 to get A^2:

111^2	$= 110^2 + 110 + 111$
	$= 12100 + 221$
	$= 12321.$

More examples and explanation

13 Squares of numbers that end with 5.

A number that ends in 5 has the form $A = 10a + 5$, where a has one digit less than A. To find the square A^2 of A, append 25 to the product $a \times (a + 1)$ of a with its successor. For example, compute 115^2. $115 = 11 \times 10 + 5$, so that $a = 11$. First compute $11 \times (11 + 1) = 11 \times 12 = 132$ (since3 = 1 + 2). Next, append 25 to the right of 132 to get 13225!

More examples and explanation

14 Product of 10a + b and 10a + c where b + c = 10.

Similar to the squaring of numbers that end with 5:

For example, compute 113×117, where $a = 11$, $b = 3$, and $c = 7$. First compute$11 \times (11 + 1) = 11 \times 12 = 132$ (since 3 = 1 + 2). Next, append 21 (= 3×7) to the right of 132 to get 13221!

More examples and explanation

15 Product of two one-digit numbers greater than 5.

This is a rule that helps remember a big part of the multiplication table. Assume you forgot the product 7×9. Do this. First find the excess of each of the multiples over 5: it's 2 for 7 (7 - 5 = 2) and 4 for 9 (9 - 5 = 4). Add them up to get 6 = 2 + 4. Now find the complements of these two numbers to 5: it's 3 for 2 (5 - 2 = 3) and 1 for 4 (5 - 4 = 1). Remember their product 3 = 3×1. Lastly, combine thus obtained two numbers (6 and 3) as 63 = 6×10 + 3.

More examples and explanation

16 Product of two 2-digit numbers.

The simplest case is when two numbers are not too far apart and their difference is even, for example, let one be 24 and the other 28. Find their average: (24 + 28)/2 = 26 and half the difference (28 - 24)/2 = 2. Subtract the squares:

$28 \times 24 = 26^2 - 2^2 = 676 - 4 = 672.$

The ancient Babylonian used a similar approach. They calculated the sum and the difference of the two numbers, subtracted their squares and divided the result by four. For example,

$$33 \times 32 = (65^2 - 1^2)/4 \\ = (4225 - 1)/4 \\ = 4224/4 \\ = 1056.$$

More examples and explanation

17 Product of numbers close to 100.

Say, you have to multiply 94 and 98. Take their differences to 100: 100 - 94 = 6 and 100 - 98 = 2. Note that 94 - 2 = 98 - 6 so that for the next step it is not important which one you use, but you'll need the result: 92. These will be the first two digits of the product. The last two are just 2×6 = 12. Therefore, 94×98 = 9212.

More examples and explanation

18 Multiplying by 11.

To multiply a 2-digit number by 11, take the sum of its digits. If it's a single digit number, just write it between the two digits. If the sum is 10 or more, do not forget to carry 1 over.

For example, 34×11 = 374 since 3 + 4 = 7. 47×11 = 517 since 4 + 7 = 11.

19 Faster subtraction.

Subtraction is often faster in two steps instead of one.

For example,

427 - 38 = (427 - 27) - (38 - 27) = 400 - 11 = 389.

A generic advice might be given as "First remove what's easy, next whatever remains". Another example:

$1049 - 187 = 1000 - (187 - 49) = 900 - 38 = 862$.

20 Faster addition.

Addition is often faster in two steps instead of one.

For example,

$487 + 38 = (487 + 13) + (38 - 13) = 500 + 25 = 525$.

A generic advice might be given as "First add what's easy, next whatever remains". Another example:

$1049 + 187 = 1100 + (187 - 51) = 1200 + 36 = 1236$.

21 Faster addition, #2.

It's often faster to add a digit at a time starting with higher digits. For example,

$$\begin{aligned} 583 + 645 &= 583 + 600 + 40 + 5 \\ &= 1183 + 40 + 5 \\ &= 1223 + 5 \\ &= 1228. \end{aligned}$$

22 Multipliply, then subtract.

When multiplying by 9, multiply by 10 instead, and then subtract the other number. For example,

$23 \times 9 = 230 - 23 = 207$.

More examples and explanation

The same applies to other numbers near those for which multiplication is simplified:

$$
\begin{aligned}
23 \times 51 &= 23 \times 50 + 23 \\
&= 2300/2 + 23 \\
&= 1150 + 23 \\
&= 1173.
\end{aligned}
$$

$$
\begin{aligned}
87 \times 48 &= 87 \times 50 - 87 \times 2 \\
&= 8700/2 - 160 - 14 \\
&= 4350 - 160 - 14 \\
&= 4190 - 14 \\
&= 4176.
\end{aligned}
$$

23 Multiplication by 9, 99, 999, etc.

There is another way to multiply fast by 9 that has an analogue for multiplication by 99, 999 and all such numbers. Let's start with the multiplication by 9.

To multiply a one digit number a by 9, first subtract 1 and form $b = a - 1$. Next, subtract b from 9: $c = 9 - b$. Then just write b and c next to each other:

$9a = bc$.

For example, find 6×9 (so that $a = 6$.) First subtract: $5 = 6 - 1$. Subtract the second time: $4 = 9 - 5$. Lastly, form the product 6×9 = 54.

24 Similarly, for a 2-digit a:

$$
\begin{aligned}
bc &= 100b + c \\
&= 100(a - 1) + (99 - (a - 1)) \\
&= 100a - 100 + 100 - a \\
&= 99a.
\end{aligned}
$$

Do try the same derivation for a three digit number. As an example,

$$
\begin{aligned}
543 \times 999 &= 1000 \times 542 + (999 - 542) \\
&= 542457.
\end{aligned}
$$

Adding a Long List of Numbers

For some people, adding a list of not very different numbers is a frequent task. To me this happens every time I fill my tax returns; there is always a few dozen professional books I buy over a year that I consider deductible. Of course, it is possible to use a spreadsheet or a specialized application, but I am accustomed to doing this the old fashioned way. Using a calculator is far from being fool proof; an innocent typo may flash a red light during an audit.

Teachers who have to compute test grade averages might also find the technique useful.

So assume we need to compute the sum of 16 numbers:

$97 + 86 + 83 + 95 + 85 + 70 + 84 + 72 + 77 + 81 + 70 + 85 + 84 + 76 + 92 + 66$.

Estimate a possible average of the numbers in the sum. For the given example, I'd choose 80. Instead of adding the given numbers, we shall add the differences of these numbers and the chosen average estimate:

$17 + 6 + 3 + 15 + 5 + (-10) + 4 + (-8) + (-3) + 1 + (-10) + 5 + 4 + (-4) + 12 + (-14)$.

At a glance, you can see that, say, $15 + 5$ cancels $(-10) + (-10)$, and 4 cancels (-4), leaving a shorter sum

$17 + 6 + 3 + 4 + (-8) + (-3) + 1 + 5 + 12 + (-14) = 30 + (-11) + 18 + (-14) = 19 + 4 = 23$.

The "dropped" part of the original sum is $80 \cdot 16 = 800 + 480 = 1280$, making the total $1280 + 23 = 1303$.

The trick here is to avoid dealing with large accumulations and a need to memorize intermediate results. Always scan the sums for possible cancellations. For example, in the last sum we may have noticed that 3 and (-3) cancel out and so do $3 + 5$ and (-8). Taking into account the latter we would get a shorter sum

$17 + 6 + 4 + (-3) + 1 + 12 + (-14)$,

in which it is hard to fail to notice that 17 and $(-3) + (-14)$ also cancel out

$17 + 6 + 4 + (-3) + 1 + 12 + (-14) = 6 + 4 + 1 + 12 = 10 + 13 = 23.$

With a little practice, such shortcuts pop into view automatically.

Multiplication by 9, 99, 999, etc.

One way to multiply a number by 9 is to multiply by 10 and then subtract the number from the product. There is another way to multiply fast by 9 and as the first one it has an analogue for multiplication by 99, 999 and all such numbers. Let's start with the multiplication by 9.

To multiply a one digit number a by 9, first subtract 1 and form $b = a - 1$. Next, subtract b from 9: $c = 9 - b$. Then just write b and c next to each other:

$9a = bc.$

For example, find 6×9 (so that $a = 6$.) First subtract: $5 = 6 - 1$. Subract the second time: $4 = 9 - 5$. Lastly, form the product 6×9 = 54.

Next, find 37×99. First, subtract 1: $36 = 37 - 1$. Then subtract $63 = 99 - 36$. Lastly, form the product: 37×99 = 3663.

Why does this work? For the multiplication by 9, $bc = 10b + c$:

$$\begin{aligned} bc &= 10b + c \\ &= 10(a - 1) + (9 - (a - 1)) \\ &= 10a - 10 + 10 - a \\ &= 9a, \end{aligned}$$

as required. Similarly, for a 2-digit a:

$$\begin{aligned} bc &= 100b + c \\ &= 100(a - 1) + (99 - (a - 1)) \\ &= 100a - 100 + 100 - a \\ &= 99a. \end{aligned}$$

Do try the same derivation for a three digit number. As an example,

$$\begin{aligned} 543 \times 999 &= 1000 \times 542 + (999 - 542) \\ &= 999 \times 542 + 999 \\ &= 999 \times 543 \end{aligned}$$

just by using the ***distributive law*** twice.

http://www.cut-the-knot.org/arithmetic/rapid/rapid.shtml

Do Super Quick Maths Calculation Using Vedic Method

Posted by geek

<u>Teaching Mental Maths Tricks to Anyone and Everyone!</u>

Learning to perform fast **mental maths** calculation will help you immensely irrespective of which field of life you deal with. Knowing these mental maths tricks will give you a positive edge over the others.Whether you are a student,aspiringengineer,statistician,scientist,school teacher or anyone else dealing with numbers,learning this quick mental tricks and techniques (popularly known as **Vedic Maths techniques**) is always going to benefit you.

You must have heard of Shakuntala Devi-the lady who performed maths calculations faster than a Computer,you can do it too, just with a little bit of practice.

For example, let say you want to **multiply 52*11**.This can be calculated in less than 1 second but if you want to do it traditionally,it will take you around 5-6 seconds.Isn't it?

SECRETS & TRICKS OF MATHEMATICS

So let see how using a simple mental maths trick,this calculation can be done in a matter of seconds...

To multiply 52 and 11,imagine there is a space between 52

52*11= 5_2 (Put an imaginary space in between)

Now,what to do with that space?

Just add 5 and 2 and put the result in the imaginary space

So, 52 * 11 =5$\underline{7}$2 (which is your answer)

Isn't it great?

Lets try some more examples:

1) 35 * 11 = 3 (3+5) 5 = 3$\underline{8}$5
2) 81 * 11 = 8 (8+1) 1 = 8$\underline{9}$1
3) 72 * 11 = 7 (7+2) 2 = 7$\underline{9}$2 etc..

With just a little bit of practice you can easily perform these simple mental maths tricks in the blink of an eye.

People sitting for competitive exams often complain that they could not complete the Question paper within a certain time period as the paper was too length(l)y.But for your information,let me tell you that all papers of all competitive exams are so designed that students can finish it within the given time period.Its just that student do not have the required efficiency.So in tight time constraint situation where time

SECRETS & TRICKS OF MATHEMATICS

plays a very important role,knowing these quick mental maths techniques will give you an edge over your competitors.It will be your X-Factor.It will give you that sharpness and smartness required to crack any competitive exams.

Lets take an example of this sum which has been taken from the 2010 question paper of a Popular Bank PO Examination (Aptitude Section):

(Q) Is 456138 divisible by 9?

Now, it only takes 2 seconds for you to determine the answer. But if you go by the traditional way then it will take you 10 seconds.So you can see the difference.Those 8 extra seconds you win,you can spend on other question.Isn't it?

No let see the solution

(Answer) To test whether a certain large number is divisible by 9 or not,'**just add all the digits of the number and if the end result is divisible by 9,then you can say that the entire large number will be divisible by 9 too**'.

$4+5+6+1+3+8=27$

Now since 27 is divisible by 9 so **456138 will be divisible by 9** too.

By now you must have some idea, how important it is to know these mental maths tricks.Knowing these simple calculation techniques gives you an advantage over others and can get you a job,get you crack any competitive exams and much more.

Here are few more mental maths tricks..

Multiply any large number by 12 mentally in seconds

To multiply any number by 12 just double last digit and thereafter double each digit and add it to its neighbour

For example **21314 * 12 = 255768**

Lets break it into simple steps:

Step 1: 021314 * 12 = _____8 (Double of Last Digit 4= 8)

Step 2: 021314 * 12 = ____68 (Now Double 1= 2, and add it to 4, 2+4=6)

Step 3: 021314 * 12= ___768 (Now Double 3=6, and add it to 1, 6+1=7)

Step 4: 021314 * 12= __5768 (Now Double 1=2, and add it to 3, 2+3=5)

Step 5: 021314 * 12= _55768 (Now Double 2=4, and add it to 1, 4+1=5)

Step 6: 021314 * 12= 255768 (Now Double 0=0, and add it to 2, 0+2=2)

SECRETS & TRICKS OF MATHEMATICS

So your final answer of **21314 * 12 = 255768**

Another example...

Calculating Square of numbers quickly...

Lets **calculate the square of 54**

So $(54)^2 = 5^2 +4 -- 4^2 = 25 +4 ----16 = 29-------16 =$ **2916**

Similarly $(55)^2 = 5^2 +5 --5^2=25+5------25=30---------25=$ **3025**

Similarly $(56)^2 = 5^2 + 6--6^2=25+6------36= 31--------36=$ **3136** etc..

Similarly try out squares of 57,58 etc..

http://mental-maths-trick.blogspot.co.uk/2011/10/do-super-quick-maths-calculation-using.html

10 Easy Arithmetic Tricks

JAMIE FRATER SEPTEMBER 17, 2007

Math can be terrifying for many people. This list will hopefully improve your general knowledge of mathematical tricks and your speed when you need to do math in your head.

1. The 11 Times Trick

We all know the trick when multiplying by ten – add 0 to the end of the number, but did you know there is an equally easy trick for multiplying a two digit number by 11? This is it:

Take the original number and imagine a space between the two digits (in this example we will use 52:

5_2

Now add the two numbers together and put them in the middle:

5_(5+2)_2

That is it – you have the answer: 572.

If the numbers in the middle add up to a 2 digit number, just insert the second number and add 1 to the first:

9_(9+9)_9

(9+1)_8_9

10_8_9

1089 – It works every time.

2. Quick Square

If you need to square a 2 digit number ending in 5, you can do so very easily with this trick. Mulitply the first digit by itself + 1, and put 25 on the end. That is all!

$25^2 = (2x(2+1))$ & 25

2 x 3 = 6

625

3. Multiply by 5

Most people memorize the 5 times tables very easily, but when you get in to larger numbers it gets more complex – or does it? This trick is super easy.

Take any number, then divide it by 2 (in other words, halve the number). If the result is whole, add a 0 at the end. If it is not, ignore the remainder and add a 5 at the end. It works everytime:

2682 x 5 = (2682 / 2) & 5 or 0

2682 / 2 = 1341 (whole number so add 0)

13410

Let's try another:

5887 x 5

2943.5 (fractional number (ignore remainder, add 5)

29435

SECRETS & TRICKS OF MATHEMATICS

4. Multiply by 9

This one is simple – to multiple any number between 1 and 9 by 9 hold both hands in front of your face – drop the finger that corresponds to the number you are multiplying (for example 9×3 – drop your third finger) – count the fingers before the dropped finger (in the case of 9×3 it is 2) then count the numbers after (in this case 7) – the answer is 27.

5. Multiply by 4

This is a very simple trick which may appear obvious to some, but to others it is not. The trick is to simply multiply by two, then multiply by two again:

58 x 4 = (58 x 2) + (58 x 2) = (116) + (116) = 232

6. Calculate a Tip

If you need to leave a 15% tip, here is the easy way to do it. Work out 10% (divide the number by 10) – then add that number to half its value and you have your answer:

15% of $25 = (10% of 25) + ((10% of 25) / 2)

$2.50 + $1.25 = $3.75

7. Tough Multiplication

If you have a large number to multiply and one of the numbers is even, you can easily subdivide to get to the answer:

32 x 125, is the same as:
16 x 250 is the same as:
8 x 500 is the same as:
4 x 1000 = 4,000

8. Dividing by 5

Dividing a large number by five is actually very simple. All you do is multiply by 2 and move the decimal point:

195 / 5

Step1: 195 * 2 = 390

Step2: Move the decimal: 39.0 or just 39

2978 / 5

step 1: 2978 * 2 = 5956

Step2: 595.6

9. Subtracting from 1,000

To subtract a large number from 1,000 you can use this basic rule: subtract all but the last number from 9, then subtract the last number from 10:

1000

-648

step1: subtract 6 from 9 = 3

step2: subtract 4 from 9 = 5

step3: subtract 8 from 10 = 2

answer: 352

10. Assorted Multiplication Rules

Multiply by 5: Multiply by 10 and divide by 2.

Multiply by 6: Sometimes multiplying by 3 and then 2 is easy.

Multiply by 9: Multiply by 10 and subtract the original number.

Multiply by 12: Multiply by 10 and add twice the original number.

Multiply by 13: Multiply by 3 and add 10 times original number.

Multiply by 14: Multiply by 7 and then multiply by 2

Multiply by 15: Multiply by 10 and add 5 times the original number, as above.

Multiply by 16: You can double four times, if you want to. Or you can multiply by 8 and then by 2.

Multiply by 17: Multiply by 7 and add 10 times original number.

Multiply by 18: Multiply by 20 and subtract twice the original number (which is obvious from the first step).

Multiply by 19: Multiply by 20 and subtract the original number.

Multiply by 24: Multiply by 8 and then multiply by 3.

Multiply by 27: Multiply by 30 and subtract 3 times the original number (which is obvious from the first step).

Multiply by 45: Multiply by 50 and subtract 5 times the original number (which is obvious from the first step).

Multiply by 90: Multiply by 9 (as above) and put a zero on the right.

Multiply by 98: Multiply by 100 and subtract twice the original number.

Multiply by 99: Multiply by 100 and subtract the original number.

Bonus: Percentages

Yanni in comment 23 gave an excellent tip for working out percentages, so I have taken the liberty of duplicating it here:

Find 7 % of 300. Sound Difficult?

Percents: First of all you need to understand the word "Percent." The first part is PER , as in 10 tricks per listverse page. PER = FOR EACH. The second part of the word is CENT, as in 100. Like Century = 100 years. 100 CENTS in 1 dollar... etc. Ok... so PERCENT = For Each 100.

So, it follows that 7 PERCENT of 100, is 7. (7 for each hundred, of only 1 hundred).

8 % of 100 = 8. 35.73% of 100 = 35.73

But how is that useful??

Back to the 7% of 300 question. 7% of the first hundred is 7. 7% of 2nd hundred is also 7, and yep, 7% of the 3rd hundred is also 7. So 7+7+7 = 21.

If 8 % of 100 is 8, it follows that 8% of 50 is half of 8 , or 4.

SECRETS & TRICKS OF MATHEMATICS

Break down every number that's asked into questions of 100, if the number is less then 100, then move the decimal point accordingly.

EXAMPLES:

8%200 = ? 8 + 8 = 16.
8%250 = ? 8 + 8 + 4 = 20.
8%25 = 2.0 (Moving the decimal back).
15%300 = 15+15+15 =45.
15%350 = 15+15+15+7.5 = 52.5

Also it's usefull to know that you can always flip percents, like 3% of 100 is the same as 100% of 3.

35% of 8 is the same as 8% of 35.

Technorati Tags: arithmetic, math

JAMIE FRATER

Jamie is the founder of Listverse. He spends his time working on the site, doing research for new lists, and cooking. He is fascinated with all things morbid and bizarre.

Figures of Fun: Amaze Your Friends With These Fantastic Maths Magic Tricks

Feb 08, 2013 10:49By **Ben Rankin**

Peter M Higgins, author and numbers man, reveals the magic behind something many of us struggle to get our heads around

Getty

A maths wizard has discovered the largest known prime number and it's an amazing 17 MILLION digits long.

Curtis Cooper, of the University of Central Missouri, in the US found it by using hundreds of computers networked together.

You might not have to be a maths genius to know that a prime is a number greater than 1, which can only be divided by itself and 1, like 2, 3, 5, 7, 11, but to find one that big, GCSE maths would be a good start.

However, in the numbers world, size doesn't matter when it comes to the fascinating and bizarre.

Here Peter M Higgins, author and numbers man, reveals the magic behind something many of us struggle to get our heads around...

Hailstone numbers

There are many simple questions about numbers that no one has been able to answer.

Start with any number - if it is even divide it by 2, if odd multiply by 3 and add 1 then keep going, writing down the sequence of numbers that you generate. For example, starting with 7 we are led by these rules through the sequence:

7 -> 22 -> 11 -> 34 -> 17 -> 52 -> 26 -> 13 -> 40 -> 20 -> 10 -> 5 -> 16 -> 8 -> 4 -> 2 -> 1.

It seems no matter what number you start with you eventually hit a 1. These sequences are called the "hailstone numbers" because, like hailstones, they go up and down a number of times before inevitably falling to Earth. However, no one has been able to prove that this has to happen every time.

Twos, threes & fives

Think of a number. Add 4, then multiply the result by 4. Subtract 8, then divide the result by 4. Finally take away your original secret number. The answer is 2.

Think of another number.

Double it. Add 9. Subtract 3. Divide by 2. Subtract your original number. The answer is 3.

Think of any three-digit number.

Add 7. Multiply by 2.

Subtract 4, then divide the result by 2.

Subtract it from the original number you thought of.

The answer is 5.

Ninety nine

Write down any two different numbers from 1 to 9. Then reverse the two numbers.

You should have two two-digit numbers.

Subtract the smaller number from the larger one.

Take the result, reverse the digits, and add that number to the one you got when you subtracted.

The answer is 99.

For example: 72 reverses to make 27.

Subtract the smaller (27) from the larger (72): 45.

Reverse these digits to make 54.

Add this to the previous number.

The answer is 99.

Threesy does it

You can discover whether a number is a multiple of 3 just by checking whether this is true for the sum of its digits.

For example, 12,894 has $1 + 2 + 8 + 9 + 4 = 24 = 3 \times 8$, so 12,894 is a multiple of 3.

You don't need to do the long division in order to find this out.

You can do this even for huge numbers that your calculator could never cope with.

For example, try: 111,222,333,444,555,666,777, 888,987. Is it divisible by 3? In fact, if you're clever, you might be able to give the answer before summing the digits.

10% up then 10% down means you lose out

A worker's boss explains that in order to stay competitive he will have to cut his pay by 10% but he will allow the employee to work 10% more hours to make up for it, "so your pay will be maintained".

Afraid not! If the worker was being paid, say, £100, the 10% cut takes him down to £90. The 10% extra hours will add back on 10% of £90, which gives him £99. He is still £1 worse off. Beware percentages - you need to know what they refer to.

Never-ending squares

Square numbers (the products of numbers multiplied by themselves) and prime numbers are important and your internet security only works because the prime numbers never run out.

You can get the endless list of squares just by adding the odd numbers up: $1 = 1 \times 1$, $1 + 3 = 4 = 2 \times 2$, $1 + 3 + 5 = 9 = 3 \times 3$, $1 + 3 + 5 + 7 = 16 = 4 \times 4$ ... and this pattern never lets you down.

However, when it comes to primes, we still have to go out hunting for them, which is why at any one time there is always a world champion largest known prime.

SECRETS & TRICKS OF MATHEMATICS

'Mind reading' trick

Choose a single-digit number, multiply it by 9 and if the answer has two digits add them together.

Subtract 5 from what you have, giving you a number. Turn the number into a letter by the rule A = 1, B = 2 and so on. Think of a country beginning with your letter. Take the last letter of your country and think of an animal that begins with that letter. It's odds on that you have a kangaroo in Denmark.

It all adds up... to 9

1x9=09 =0+9=9

2x9=18 =1+8=9

3x9=27 =2+7=9

4x9=36 =3+6=9

5x9=45 =4+5=9

6x9=54 =5+4=9

7x9=63 =6+3=9

8x9=72 =7+2=9

10 x 9 = 90 = 9 + 0 = 9

Magic cube

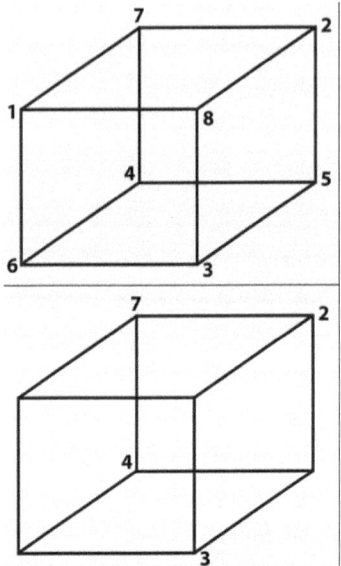

Think you can conjure the right numbers? Try this little trick then.

Label the corners of the cube on the left with different numbers in the range 1 to 8 so that each face adds up to the same total.

The solution is on the right.

One, two, three

1 x 1 = 1
11 x 11 = 121
111 x 111 = 12321
1111 x 1111 = 1234321
11111 x 11111 = 123454321

111111 x 111111 = 12345654321
1111111 x 1111111 = 1234567654321
11111111 x 11111111 = 123456787654321
111111111 x 111111111 = 12345678987654321
1 x 9 + 2 = 11
12 x 9 + 3 = 111
123 x 9 + 4 = 1111
1234 x 9 + 5 = 11111
12345 x 9 + 6 = 111111
123456 x 9 + 7 = 1111111
1234567 x 9 + 8 = 11111111
12345678 x 9 + 9 = 111111111
123456789 x 9 +10 = 1111111111

A final thought

Averages can be misleading - the average human has one breast and one testicle.

http://www.mirror.co.uk/news/weird-news/amaze-your-friends-with-these-fantastic-maths-1593633

Use these calculator tricks to impress and astound your friends!

Is That Your Final Answer?

1. Have someone pick a number between 1 and 9.
2. Now have him use a calculator to first multiply it by 9, and then multiply it by 12,345,679 (notice there is no 8 in that number!).
3. Have the person show you the result so you can tell him the original number he selected! How? Easy. If he selected 5, the final answer is 555,555,555. If he selected 3, the final answer is 333,333,333. The reason: 9 x 12345679 = 111111111. You multiplied your digit by 111111111. *(By the way, that 8-digit number (12,345,679) is easily memorized: only the 8 is missing from the sequence.)*

The 421 Loop

1. Pick a whole number and enter it into your calculator.

2. If it is even, divide by 2. If it is odd, multiply by 3 and add 1.
3. Repeat the process with the new number over and over. What happens?
4. The sequence *always* ends in the «loop»: 4.....2.....1.....4.....2.....1...

Example: Start with 13.

13 is odd, so we multiply by 3 and add 1. We get 40. (13 x 3 = 39 + 1= 40)
40 is even, so we divide by 2. We get 20. (40 / 2 = 20)
20 is even, so we divide by 2 and get 10.
10 is also even so we divide by 2 again and get 5.
5 is odd so we multiply by 3 and add 1. We get 16.
16 is even, so we divide by 2 and get 8.
8 is also even so we divide by 2 again and get 4.
4 is even so we divide by 2. We get 2.
2 is even, so we divide by 1 and get 1.
1 is odd, so we multiply by 3 and add 1. We get 4.
4 is even so we divide by 2. We get 2. And so we begin the loop 4.....2.....1.....4.....2.....1...

Good Luck or Bad Luck?

1. Have someone secretly select a three-digit number and enter it twice into her calculator. (For example: 123123) Have her concentrate on the display. You will try to discern her thoughts!

2. From across the room (or over the phone), announce that the number is divisible by 11. Have her verify it by dividing by 11.
3. Announce that the result is also divisible by 13. Have her verify it.
4. Have him divide by his original three-digit number.
5. Announce that the final answer is 7.

You can use this to predict Good Luck for him. If you wish to predict Bad Luck, have him divide by 7 in step 3; the final answer will be 13.

Why does this work? Entering a three-digit number twice (123123) is equivalent to multiplying it by 1001. (123 x 1001 = 123,123). Since 1001 = 7 x 11 x 13, the six-digit number will be divisible by 7, 11, 13, and the original three-digit number.

The Secret of 73

1. For this trick, secretly write 73 on a piece of paper, fold it up, and give to an unsuspecting friend.
2. Now have your friend select a four-digit number and enter it twice into a calculator. (For example: 12341234)
3. Announce that the number is divisible by 137 and have him verify it on his calculator.
4. Next, announce that he can now divide by his original four-digit number. After he has done so, dramatically command him to look at your prediction on the paper. It will match his calculator display: 73!

Why does this work? Entering a four-digit number twice (12341234) is equivalent to multiplying it by 10001. (1234 x 10001 = 12341234). Since 10001 = 73 x 137, the eight-digit number will be divisible by 73, 137, and the original four-digit number.

The 6174 loop

1. Select a four-digit number. (Do not use 1111, 2222, etc.)
2. Arrange the digits in increasing order.
3. Arrange the digits in decreasing order.
4. Subtract the smaller number from the larger number.
5. Repeat steps 2, 3, and 4 with the result, and so on. What happens?

Let's try 7173

Arrange the digits in increasing order. 1377
Arrange the digits in decreasing order. 7731
Subtract the smaller number from the larger number. 7731 - 1377 = 6354
Repeat the process with 6354
6543 - 3456 = 3087
8730 - 0378 = 8352
8532 - 2358 = 6174
7641 - 1467 = 6174
7641 - 1467 = 6174
7641 - 1467 = 6174 (we're in a loop!)

Amazingly, all four-digit numbers (not multiples of 1111) end up in the 6174-loop. No reason has been found for this phenomenon.

The Golden Prediction

This trick takes considerable time, but the effect is spectacular.

Give someone a sheet of paper and a pencil and tell him to:

1. Number the first 25 lines (1, 2, 3,...).
2. Write any two whole numbers on the first two lines.
3. Add the two numbers and write the sum on the third line.
4. Add the last two numbers and write the sum on the next line.
5. Continue this process (add the last two, write the sum) until he has 25 numbers on his list.
6. Select any number among the last five on his list, and divide by the previous number (the number above it). Now for the trick!

Remind him that you do not know his original two numbers or any of the 25 numbers, that you do not know which of the 25 numbers he selected right now, and therefore you cannot possibly know the number on the display.

With great concentration and much difficulty, you divine the number presently on his calculator: "I'm getting a One... then something funny - oh! a decimal point! Then... a Six..another One.. and an Eight, I think.. Now I'm getting a blank..nothing...

Oh! It's a Zero!..then a Three... and... another Three?... then a Nine... had enough?"

That's right! If your subject selects any number between the last five (#21 through #25) and divides it by the number above it, **he'll always get 1.618033989...**, which just happens to be theGolden Mean! *(provided, of course, he did all the addition correctly in steps 3-5 above)*

Why does this work? It's an incredible bit of mathematical trivia. Begin with any two whole numbers, make a Fibonacci-type addition list, take the ratio of two consecutive entries, and the ratio approaches the Golden Mean! The further out we go, the more accurate it becomes. That's why we need 25 numbers: to obtain sufficient accuracy. The proof requires familiarity with the Fibonacci Sequence, pages of algebra, and a knowledge of Limits, all of which go far beyond the scope of this site.

Interesting fact: if you divide one of your last 5 numbers by the *next* number (instead of the previous number), the result is the same decimal without the leading 1. (0.618033989)

http://www.maths.dit.ie/scienceweek/calculatortricks.htm

Weird Mathematical Tricks That Shouldn't Work - But Do

By **David Baugher**

At one time or another everyone gets stymied by a numbers problem. It seems to add up but you just don't quite know why since it doesn't feel like it should. But that is the joy of some numerical brain teasers. It can be frustratingly fun to find quirky concepts that don't seem like they should work yet somehow they still do. Below are a few weird mathematical tricks and conundrums that you won't believe but if you do the math, the figures do indeed total up in ways that appear to defy common sense.

The stock oddity

Some folks get a thrill from playing the market. Maybe you are one of them. Sometimes you win. Sometimes you lose but those are the risks of the game.

What's the oddity?

Winning and losing aren't as simple as you think.

Let's say you invest $10 in the market and you make a 10 percent return. You now have $11. Now, let's say you lose 10 percent. Out of $11, that's $1.10 leaving you with $9.90 which means you are down ten cents on the deal. You gained the same percentage as you lost yet you came out behind.

Well, you might speculate it has to do with the order of the transaction. After all, the 10 percent you lost was bigger than the 10 percent you gained because you were already up on the deal. That means reversing the order should have the opposite effect. Right?

Start with $10. Now lose 10 percent first. You have nine dollars. Then gain ten percent. That's 90 cents leaving you with…$9.90.

Yep. You lost money again.

Strange as it may seem, a gain and a loss of the exact same percentage will always leave you with less cash - regardless of the order in which they occur.

The casino fallacy

Casinos can be more fun than the stock market but the element of chance makes them notoriously risky hangouts. But is there a way to beat the odds? At one game, you can - well, sort of.

That game is roulette where you can bet on red or black to win. How can you guarantee a win? Well, just put a dollar on black. If it loses, put down two dollars. A win will give you back your original dollar plus one more. But if it loses, you are now out three dollars. So put down four. A win will put you ahead by a dollar. In fact, in an infinite series, you can indeed guarantee a win that puts you ahead if you just keep doubling your previous bet again with each loss.

Why don't people do this? Because in the real world it is very time consuming and, worse, could be very hazardous to your finances. That's because the doubled bets can get very large very quickly. If you hit a streak of a dozen reds in a row, you will be betting thousands of dollars just to make one and such streaks, known to statisticians as a run, can happen. If you run out of money on, say, the 15th spin, you stand to lose some $40,000.

And if you win? Well, you are ahead by a single dollar. Not a very good way to make cash. Theoretically, of course, it could work, however slowly, with a very patient, low-stakes casino and an infinite amount of cash but if you have that, why are you gambling in the first place?

Overall, it is not a recommended method for making a buck.

DID YOU KNOW?

Every even integer higher than two can be arrived at by the sum of two prime numbers (figures which are divisible evenly only by themselves and one). Known as Goldbach's Conjecture, this theory has been shown to be true going up to very high numbers though it has never been conclusively proven. Still, no number has been located that contradicts this strange rule.

Three men and a hotel room

Here's a brain teaser. Three men rent a hotel room which costs $30. But after they've checked in, the manager discovers he's made an error. The room was only $25. He goes up with five ones to give them. The men are appreciative of the refund but to make things simpler they each take a dollar and tip him the remaining two bucks. Since each man received a dollar back, they've now only paid $9 each for the room, not $10. Yet, if we add 9+9+9 we get $27. Then there is the two the manager received. That makes $29. Yet the total was originally $30.

Where is the other dollar?

The trick to this is to simply understand the question better. There really is no missing dollar. Twenty-five went for the room, three for the men and two for the manager. That equals $30. Adding it up the other way is simply obfuscation. Between the tip and the room, the men spent $27. They got three back. That's $30.

Three-card monte

Let's say someone shows you three cards face down - a red and two blacks. If you pick the red, you win the game. You begin by selecting a card but you don't look at it. The dealer then removes one of the two remaining cards from the game. (Obviously, he won't remove the red, if he has it, or the game is over.) You are now given a choice: Stay with your original card or switch to the new one. Which should you choose?

Most people say it doesn't matter. Since you have seen neither card, the odds are even.

Surprisingly, however, that is incorrect. Incredibly, switching to the other card doubles your chances of holding the red. This seems counterintuitive since you have seen none of the cards and picked the original at random. How could switching from one random card to another help? The answer lies in a statistical oddity. Your original chance of picking the correct card was one-in-three. That means there was a two-in-three chance you did not pick the red and it is one of the other two cards. When the dealer removes one of those from the game, he cannot remove the red so the card he dumped must be black. But that means the odds don't change. Since there was a two-in-three chance one of the two cards he held was the red and a 100% chance the card he threw away was black, there is still a two-in-three chance the remaining card is the correct one. Strange as it seems, the odds between the two cards are not even.

The magic of 11

Times tables were the grueling displeasure of every school kid and multiples involving two figures with a tens column were more depressing still. Most of us need pen and paper for multiplications involving a pair of double-digit numbers.

Yet this isn't true with the number 11 if you use one weird trick that doesn't seem like it should work but does. Take the two digits of the figure you are multiplying by 11 and add them. Then put the one-digit sum in between them. Example: 34x11. 3+4=7. The answer is 374.

If the sum is two-digits, put the second digit between the two numbers and add one to the original number's tens column. Example: 11x69. 6+9=15. The five goes in-between and add one to the six. The answer is 759.

Contrary to popular belief, math can indeed be enjoyable. In fact, trying to figure out conundrums like these can be a nice way to give your mind a bit of exercise. Considering things that look like they shouldn't add up but do can allow us to confuse friends, stimulate conversation at parties or just provide an interesting way to pass a lazy afternoon in thought with a few weird math tricks.

http://weird.answers.com/facts/weird-mathematical-tricks-that-shouldnt-work-but-do

NOT STATISTICALLY SIGNIFICANT AND OTHER STATISTICAL TRICKS.

by Hang
Not statistically significant…

Most people have no idea what "Not statistically significant" means and I don't see the media being too eager to fix this.

Say you read the following piece in a newspaper:

A study done at the University of Washington showed that, after controlling for race and socioeconomic class, there was no statistically significant difference in athletic performance between those who stretched for 5 minutes before running and those who did no stretching at all.

What do you conclude from that? Stretching is useless? WRONG.

Here's what the hypothetical study actually was: I picked four random guys on campus and asked two of them to stretch and two of them not to. The ones who stretched ran 10% faster.

Why is this then not statistically significant? Because the sample size was too small to infer anything useful and the study was designed poorly.

All "not statistically significant" tells you is that you can't infer anything from the study but word the study carefully enough and you can have people believe the opposite is true.

Have you ever heard the claim "There's no statistically significant difference between going to an elite Ivy League school and an equally good state school?"

Well, from **this paper** (via a comment in an **Overcoming Bias post**):

For instance, Dale and Krueger (1999) attempted to estimate the return to attending specific colleges in the College and Beyond data. They assigned individual students to a "cell" based on the colleges to which they are admitted. Within a cell, they compared those who attend a more selective college (the treatment group) to those who attended a less selective college (the control group). If this procedure had gone as planned, all students within a cell would have had the same menu of colleges and would have been arguably equal in aptitude. The procedure did not work in practice because the number of students who reported more than one college in their menu was very small. Moreover, among the students who reported more than one college, there was a very strong tendency to report the college they attended plus one less selective college. Thus, there was almost no variation within cells if the cells were based on actual colleges. Dale and Krueger were forced to merge colleges into crude "group colleges" to form the cells. However, the crude cells made it implausible that all students within a cell were equal in aptitude, and this implausibility eliminated the usefulness of their procedure. Because the procedure works best when students have large menus and most student do not have such menus, the procedure essentially throws away much of the data. A procedure is not good if it throws away much of the data and still does not deliver "treatment" and "control" groups that are plausibly equal in aptitude. Put another way, it is not useful to discard good variation in data without a more than commensurate reduction in the problematic variation in the data. In the end, Dale and Krueger predictably generate statistically insignificant results, which have been unfortunately misinterpreted by commentators who do not sufficient econometric knowledge to understand the study's methods.

In other words, the study says no such thing, it simply says the study itself was not sufficient to prove that Ivy League educations made you more money because the data wasn't good enough and yet the media has twisted this into a positive assertion that state schools do indeed make you as much money as Ivy Leagues.

I'm generously inclined to believe that most cases that I see of this error are caused by incompetence but it's pretty trivial to see how this could be used for malice. Want the public to believe that Internet usage doesn't cause social maladjustment? Just design a shitty study and claim "We found no statistical difference in social competence between heavy internet users, light internet users and non users". Bam, half the PR work has already been done for you.

Controlling for...

Here's another statistical gem I see all the time:

An analysis done at the University of Washington showed that there was zero correlation between race and financial attainment after controlling for IQ, education levels, socioeconomic status and gender.

Heartwarming right, it means if we put blacks and whites in the same situation, they should earn the same amount of money. WRONG.

The key here is to see that we're looking for financial attainment and controlling for socioeconomic status. Those two things mean the same damn thing. Basically, all this study told us was that being rich causes you to be rich.

Most people view the "controlling for" section of statistical reporting as a sort of benign safeguard. Controlling for things is like… due diligence right, the more the better… It's easy to numb people into a hypnotic lull with a list of all the things you control for.

But controlling for factors means you get to hide the true cause for things under benign labels. That's why I'm always so wary of studies that control for socioeconomic status or education levels, especially when they don't have to. Sure, socioeconomic status might cause obesity but what causes socioeconomic status.

Conclusion

When people do bother to talk about statistical manipulation, they usually focus on issues of statistical fact: Aggressive pruning of outliers, shotgun hypothesis testing and overly loose regressions. But why bother with having to sneak poorly designed studies past peer review when you can just publish a factually accurate study which implies a conclusion completely at odds with the data? That way, you sneak past the defenses of anyone who actually does know something about statistics.

Sometimes, I swear, the more statistically savvy a person thinks they are, the easier they are to manipulate. Give me a person who mindlessly parrots "Correlation does not imply causation" and I can make him believe any damn thing I want.

http://blog.figuringshitout.com/not-statistically-significant-and-other-statistical-tricks/

How to Lie and Cheat with Statistics

Ok...this is what you have been waiting for. How can you lie with statistics? Actually, the purpose of this page is NOT to teach you how to lie and cheat with statistics. Rather, I hope you will learn how it is possible to be misled and how to spot "statistical abuse." You can find poor use of statistics everywhere...magazines, newspapers, polls, TV, even research papers. I do not want to hear of any of you readers using these poor methods.

The Average Switcheroo

Which average (mean, median, or mode) should be used to report the results of an experiment or survey? All three types of averages describe the data truthfully. However, depending on the data, the mean, median and mode can be very different from one another.

Here is an example: suppose you asked 7 people how much money they brought to school. Here are the answers:

Person	Money
John	2
Ann	3
Bob	1
Mary	10
Sue	5
Carol	2
Ken	999

What is the mean, median and mode of the amount of money brought to school?

Mean:

Median:

Mode:

Bottom of Form If you use the mean as the average, then it will look like people bring a lot of money to school. However, if you use the median and mode, it will look like people do not bring much money to school. Each way to describe the numbers is correct. However, because "Ken" brought $999 to school, the mean is much different than the median and mode. Therefore, when you hear someone say, "The **average**...", make sure you know *which* type of average (mean? median? mode?) they are talking about.

SECRETS & TRICKS OF MATHEMATICS

The Meaningless Mean

QUESTION: When is a mean meaningless?

ANSWER: When a mean is created from ordinal data.

You will find this trick in many places...unfortunately this error can sometimes be found in research papers.

If you remember from the page on scales, ordinal data can be ranked, but nothing can be said about differences between numbers. Let's use the hot pepper example again. A hot pepper is scored as a "1", a hotter pepper is scored as a "2" and the hottest pepper is scored as a "3." Let's say you wanted to test the hotness of these 3 peppers and gave them to people to taste. Here are the results:

Person	Pepper A	Pepper B	Pepper C
John	2	1	3
Mary	1	2	3
Rob	2	1	3
Sue	1	2	3
Ann	2	1	3

What can be said and what cannot be said about the taste of these peppers? First, all of the people thought that Pepper C was the hottest. It also looks likes Peppers A and B tasted about the same. If you took the means of these numbers you would get:

Mean hotness of Pepper A = 1.6

Mean hotness of Pepper B = 1.4

Mean hotness of Pepper C = 3.0

But is this fair? Can you say that Pepper C was about twice as hot as Peppers A and B? Probably not. Here's why. What if Pepper A and Pepper B were not very hot at all, but Pepper C was so hot that you had to drink many glasses of water to cool the taste. The numerical differences between the taste of these peppers has no meaning. The ranking of the peppers is fine...Pepper C IS hotter than Pepper A and B, but these data give NO indication of HOW MUCH hotter. Therefore, be careful when you read about differences between numbers that come from rankings.

Actually, it is not even correct to create a mean from these data. The mean hotness of one hot pepper (score = 1) and one of the hottest peppers (score = 3) does NOT necessarily give you the score of a hotter pepper (score = 2). It may be that the hottest pepper is 100 times as hot as the least hot pepper. These data just do NOT give you this information. They only give you rankings.

The Sampling Trick

It is essential that data come from a random sample of the population. If it doesn't, then the results of the experiment or survey may not be an accurate reflection of the population. This happened in the early 1900s when polls were taken during the U.S. presidential campaign of Franklin D. Roosevelt (FDR).

The polls surveyed only those people with telephones. The pollsters predicted one candidate would win, but FDR actually won the real election. The poll did NOT accurately reflect all of the voters because the opinions of only one part of the population (wealthy people with telephones) were taken into account.

The size of the sample is another important consideration. If you flipped a coin 5 times and it came up "heads" 4 times, would you be correct to say that the coin will land on heads 80% of the time? It did for your sample of 5 flips. But what would happen if you flipped the coin 100 times or 1000 times. Would heads still come up 80% of the time? The number of people or number of trials in an experiment that are needed to give you an accurate estimate of the population is dependent on several variables. One important consideration is how much variability there is in the response. If a response has a high degree of variability, then a larger sample will be needed. In general, the larger the sample size, the better the estimation.

Games with Graphics

Misusing and abusing graphics are easy ways to mislead people. People like to see graphs for a quick way to evaluate a set of numbers. But BEWARE! Make sure you are not fooled. Let's use pumpkins grown in the gardens of Mary, Joe and Ann. Here is the first graph:

SECRETS & TRICKS OF MATHEMATICS

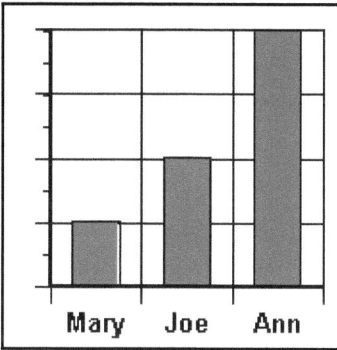

This graph does not say very much because there is no scale on the y-axis. Does the graph represent the weight, volume, width or height of the pumpkins? It does not say.

Here is a graph that is much better:

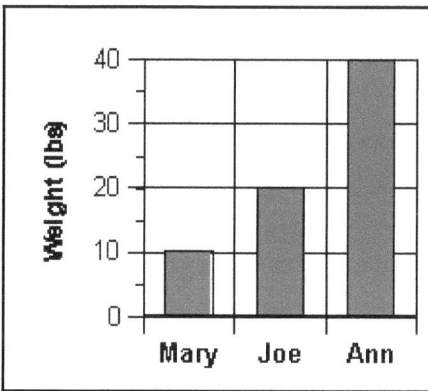

Now we know that the graph refers to the weight of the pumpkins and we know how much each pumpkin weighs because the numbers are given. This is a fair graph.

What if you wanted to convince people that Ann's pumpkin was bigger than Mary's and Joe's pumpkin. Look at this graph:

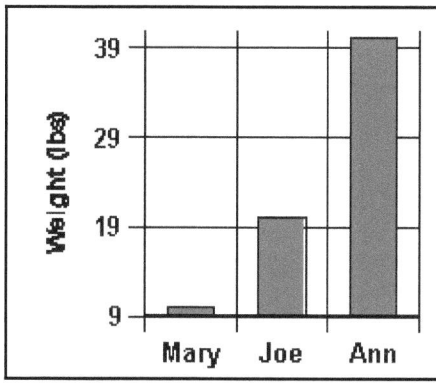

The numbers are the same, but the y-axis has been changed. Now it *appears* that Ann's pumpkin is much bigger than the other two.

SECRETS & TRICKS OF MATHEMATICS

What if you wanted to convince people that all the pumpkins were about the same size. Look at this graph:

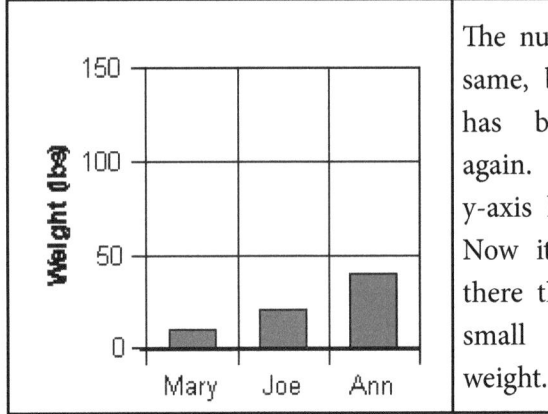

The numbers are the same, but the y-axis has been changed again. This time the y-axis has expanded. Now it *appears* that there there is only a small difference in weight.

Often a picture is used to represent data. Here is a fair way to show the difference in the weights of the pumpkins using a picture:

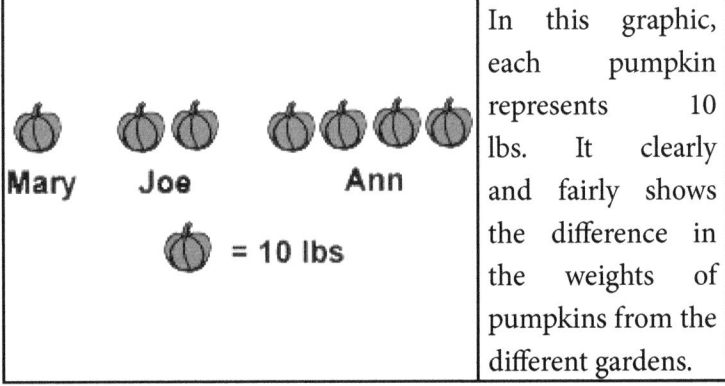

In this graphic, each pumpkin represents 10 lbs. It clearly and fairly shows the difference in the weights of pumpkins from the different gardens.

However, let's try to show that Ann's pumpkin is much bigger than the rest:

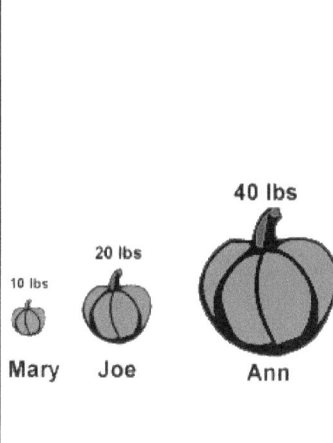

This graphic distorts the data. To show the differences in the weights, this picture changes the height of each pumpkin to represent pumpkin weight. Joe's pumpkin (20 lbs) is twice as high as Mary's (10 lbs.); Ann's (40 lbs.) is twice as high as Joe's (20 lbs.) and four times as high as Mary's (10 lbs). Is this fair? I think not! The reason is because as the height of the pumpkin is increased, the WIDTH of the pumpkin increased. Therefore, while the heights are in proportion the AREAS of the pumpkins are not. Remember, the formulas to determine area:

Area of a rectangle = Height X Width

Area of a circle = pr^2

So this picture makes it look like Ann's pumpkin is much larger than Mary's and Joe's. I also used different sized letters for the different pumpkin weights to give the impression that Ann's pumpkin was larger.

SECRETS & TRICKS OF MATHEMATICS

Meaningless Graphics

Newspapers and magazines like to use colorful pictures to represent public opinion and survey responses. However, often times the pictures are too simple to give meaningful information. Take this example:

This map shows how people in different states of the US like pizza. (I just made up these data). The code for the state color is:

Red States = People Love Pizza

Yellow States = People Like Pizza

Purple States = People Hate Pizza

That's all the information we have. The map really doesn't say very much. We don't know how it was determined that people like pizza...were people asked if they liked pizza? Were people asked how much pizza they ate in a week? a month? a year? Was the number of pizzas purchased at stores in different states counted? Was the number of pizza restaurants in different states counted?

We also do not know if there are any real differences between how much people like pizza in the different states. How much do people love pizza in California? What is the difference between how much people love pizza in Utah compared to how much they like pizza in Nevada? There are no scales or measurements to indicate any of this information. Although this type of graphic gives almost no information, it is used frequently in many popular magazines.

For more ways to misuse statistics, there are two interesting books:

Darrell Huff, *How to Lie with Statistics*, W.W. Norton & Co., New York, 1954 (reissued in 1982 and 1993).

Cooper B. Holmes, *The Honest Truth About Lying With Statistics*, Charles C. Thomas, Springfield, 1990.

Weird Statistics

DECEMBER 28, 2012 MRX

- » In chess, there are 169,518,829,100,544,000,000,000,000,000 ways to play the first ten moves.
- » It only takes 7 pounds of pressure to rip your ear off.
- » $26 billion in ransom has been paid out in the U.S. in the past 20 years.
- » You use more calories eating celery than there are in the celery itself.
- » On average, there are 178 sesame seeds on each McDonalds BigMac bun.
- » There are 1 million ants for every person in the world.
- » Odds of being killed by a dog – 1 in 700,000.
- » Odds of dying while in the bath tub – 1 in 1 million.
- » Odds of being killed by space debris – 1 in 5 billion.
- » Odds of being killed by poisoning – 1 in 86,000.
- » Odds of being killed by freezing – 1 in 3 million.
- » Odds of being killed by lightening – 1 in 2 million.
- » Odds of being killed in a car crash – 1 in 5,000.
- » Odds of being killed in a tornado – 1 in 2 million.
- » Odds of being killed by falling out of bed – 1 in 2 million.
- » Odds of being killed in a plane crash -1 in 25 million.
- » If you played all of the Beatles' singles and albums that came out between 1962 and 1970 back to back, it would only last for 10 hours and 33 minutes.

- Termites eat through wood 2 times faster when listening to rock music.
- The Apollo 11 only had 20 seconds of fuel when it landed.
- 13 people are killed each year by vending machine's falling on them.
- There is a 1/4 pound of salt in every gallon of seawater.
- About 1/3 of American adults are at least 20% above their recommended weight.
- The average talker sprays about 300 microscopic saliva droplets per minute, about 2.5 droplets per word.
- The average smell weighs 760 nanograms.
- The Earth experiences 50,000 earthquakes each year.
- Skin temperature does not go much above 95 degrees even on the hottest days.
- 314 Americans had buttock lift surgery in 1994.
- Annual growth of WWW traffic is 314,000%.
- Experts at Intel say that microprocessor speed will double every 18 months for at least 10 years.
- The Earth's revolution time increases .0001 seconds annually.
- Driving at 75 miles (121 km) per hour, it would take 258 days to drive around one of Saturn's rings.
- Driving 55 miles (88 km) per hour instead of 65 miles (105 km) per hour increases your car mileage by about 15%.
- Airbags explode at 200 miles (322 km) per hour.
- If we had the same mortality rate now as in 1900, more than half the people in the world today would not be alive.
- 1/3 of all cancers are sun related.

SECRETS & TRICKS OF MATHEMATICS

- » The average person flexes the joints in their finger 24 million times during a lifetime.
- » There are more than 1,000 chemicals in a cup of coffee.
- » It would take 7 billion particles of fog to fill a teaspoon.
- » The average iceberg weighs 20 million tons.

http://all-funny.info/weird-statistics